W0106942

Targeting of Drugs 2
Optimization Strategies

NATO ASI Series

Advanced Science Institutes Series

A series presenting the results of activities sponsored by the NATO Science Committee, which aims at the dissemination of advanced scientific and technological knowledge, with a view to strengthening links between scientific communities.

The series is published by an international board of publishers in conjunction with the NATO Scientific Affairs Division

A	**Life Sciences**	Plenum Publishing Corporation
B	**Physics**	New York and London
C	**Mathematical and Physical Sciences**	Kluwer Academic Publishers
		Dordrecht, Boston, and London
D	**Behavioral and Social Sciences**	
E	**Applied Sciences**	
F	**Computer and Systems Sciences**	Springer-Verlag
G	**Ecological Sciences**	Berlin, Heidelberg, New York, London,
H	**Cell Biology**	Paris, and Tokyo

Recent Volumes in this Series

Series A: Life Sciences

Targeting of Drugs 2
Optimization Strategies

Edited by

Gregory Gregoriadis

School of Pharmacy
University of London
London, United Kingdom

Anthony C. Allison

Syntex Research
Palo Alto, California

and

George Poste

SmithKline Beecham Pharmaceuticals
King of Prussia, Pennsylvania

Plenum Press
New York and London
Published in cooperation with NATO Scientific Affairs Division

Proceedings of a NATO Advanced Study Institute
on Targeting of Drugs: Optimization Strategies,
held June 24–July 5, 1989,
in Cape Sounion, Greece

Library of Congress Cataloging-in-Publication Data

NATO Advanced Study Institute on Targeting of Drugs: Optimization
  Strategies (1989 : Ákra Soúnion, Greece)
    Targeting of drugs 2 : optimization strategies / edited by Gregory
  Gregoriadis, Anthony C. Allison, and George Poste.
        p.   cm. -- (NATO ASI series. Series A, Life sciences ; v.
  199)
    "Proceedings of a NATO Advanced Study Institute on Targeting of
  Drugs: Optimization Strategies, held June 24–July 5, 1989, In Cape
  Sounion, Greece"--T.p. verso.
    "Published in cooperation with NATO Scientific Affairs Division."
  Includes bibliographical references.
  ISBN 978-1-4684-9003-9   ISBN 978-1-4684-9001-5 (eBook)
  DOI 10.1007/978-1-4684-9001-5

    1. Drug targeting--Congresses.   I. Gregoriadis, Gregory.
  II. Allison, Anthony C. (Anthony Clifford), 1925-   . III. Poste,
  George. IV. North Atlantic Treaty Organization. Scientific Affairs
  Division. V. Title. VI. Title: Targeting of drugs two.
  VII. Series.
    [DNLM: 1. Drugs--administration & orgainzation--congresses.   QV
  785 N279t 1989]
  RM301.63.N373   1989
  615'.7--dc20
  DNLM/DLC
  for Library of Congress   90-14327
                                                          CIP

© 1990 Plenum Press, New York
Softcover reprint of the hardcover 1st edition 1990

A Division of Plenum Publishing Corporation
233 Spring Street, New York, N.Y. 10013

All rights reserved

No part of this book may be reproduced, stored in a retrieval system, or transmitted
in any form or by any means, electronic, mechanical, photocopying, microfilming,
recording, or otherwise, without written permission from the Publisher

PREFACE

The NATO Advanced Studies Institute series "Targeting of Drugs" was originated in 1981. It is now a major international forum, held every two years in Cape Sounion, Greece, in which the present and the future of this important area of research in drug carriers is discussed in great depth. Four previous ASIs of the series dealt with drug carriers of natural and synthetic origin, their interaction with the biological milieu and with ways by which the latter influences such interaction. The present book contains the proceedings of the 5th NATO ASI "Targeting of Drugs: Optimization Strategies" held in Cape Sounion during 24 June–5 July, 1989. A logical sequel to the last one, the ASI deals with strategies by which milieu interference curtailing the function of drug carriers is circumvented or removed.

We express our appreciation to Drs. R. Langer and E. Tomlinson for their valuable advice throughout the planning of the ASI and to Dr. G. Deliconstantinos who, as Chairman of the Local Committee contributed so effectively to its success. The ASI was held under the sponsorship of NATO Scientific Affairs Division and co-sponsored and generously financed by Smith Kline French Laboratories (now SmithKline Beecham), Philadelphia, USA. Financial assistance was also provided by CIBA Geigy (Horsham), Schering (West Berlin), Farmitalia Carlo Erba (Milan), Liposome Technology Inc. (Menlo Park), Pfizer (Sandwich), Dior (Paris), Syntex Research (Palo Alto), ICI Pharmaceuticals (Mereside), Boehringer (Mannheim), Wyeth (Taplow), Merck Sharp Dohme (Rahway), Sandoz A.G. (Basle) and Lilly Research Centre Ltd. (Windlesham).

Gregory Gregoriadis
Anthony C. Allison
George Poste

---

1. Targeting of Drugs (eds. G. Gregoriadis, J. Senior, J. and A. Trouet), Plenum, 1982.
2. Receptor-Mediated Targeting of Drugs (eds. G. Gregoriadis, G. Poste, J. Senior, and A. Trouet), Plenum, 1984.
3. Targeting of Drugs with Synthetic Systems (eds. G. Gregoriadis J. Senior and G. Poste), Plenum, 1986.
4. Targeting of Drugs: Anatomical and Physiological Considerations (eds. G. Gregoriadis and G. Poste), Plenum, 1988.

CONTENTS

BIOLOGICAL DISPERSION AND THE DESIGN OF SITE-SPECIFIC PROTEIN THERAPEUTIC SYSTEMS

E. Tomlinson

Advanced Drug Delivery Research, Ciba-Geigy
Pharmaceuticals, Horsham, West Sussex RH12 4AB, UK

INTRODUCTION

A drug acts when it reaches its pharmacological site of action.
However, ideal clinical effectiveness relies additionally on the right
amount of drug reaching its site(s) of action at the right rate and
frequency, on (often) a drug-free interval (at the receptor site), and on
not critically interacting with non-target sites. Few drugs in use today
attain such ideality. As more attention is paid to drug/receptor interactions, (often through the use of molecular modelling procedures and/or
the use of cloned endogenous proteins which can act as templates for the
designed fit of agonist or antagonist drugs), increased effort is being
focussed on controlling the biological dispersion of drugs. This activity,
which was once reserved almost exclusively for cytotoxic drugs, is an
increasingly important aspect of the discovery and drug development process, particularly in the design of therapeutic proteins. Approaches to
site-specific delivery include simple low molecular weight prodrugs activated at sites of disease, suicide enzyme substrate inhibitors; polymeric
soluble and particulate macromolecular carriers; and unique site-specific
therapeutic proteins.

Site-specific drugs and carriers are often trivially referred to as
targeting systems or magic bullets. The magic bullet concept is a poor one,
since it conjures up the image of an aimed missile have a defined and pre-
ordained trajectory. However, in reality, site-selectivity relies on the
drug or carrier having a higher affinity for a particular feature of the
normal or diseased body. Conceptually, it is more a case of "I don't know
where I'm going to, but I'll know when I get there".

THERAPEUTIC PROTEIN SYSTEMS

Many polypeptides and proteins having unique pharmacological properties
are expected to be used clinically in the coming decades. These usually
have regulatory or homeostatic functions, and include both endogenous polypeptides and proteins, and their (heterologous) derivatives. This latter
class of molecules may be produced by _inter alia_, site-directed mutagenesis,
proteolysis, ligated gene fusion, protein aggregation and/or conjugation
with (other) biologically active effector functions (Tomlinson, 1989).
Proteins can be considered as drugs (for example, drug/antibody complexes).

Targeting of Drugs, Edited by G. Gregoriadis _et al._
Plenum Press, New York, 1990

Many of the proteins proposed for therapeutic use are glycoproteins whose biological disposition is primarily due to three properties: namely, their chemical and metabolic stability, size and shape, and surface features. The biological half-life of most polypeptides and proteins is short, due generally to a poor chemical stability, and/or rapid liver metabolism and kidney excretion. Also, the instability of paracrine- and autocrine-like mediators is largely due to their degradation by peptidases and proteinases in the vascular endothelium, liver, and kidneys, etc. Further, the terminal amino acids of proteins may serve to control their intracellular metabolic stability (and hence their intracellular residence time). Protein-engineering methods may be used to replace labile amino acids, with, for example, the oxidation-resistant amino acids alanine, serine or threonine, or to produce proteins having differing foldings, potentially leading to proteins which are protected from inactivation.

Pharmacology

For most therapeutic proteins the relation between applied dose and effect is highly critical, particularly as non-linear dose-effect relationships are often found (e.g. with parathyroid hormone, substance P and δ-sleep inducing peptide). As we and others, notably George Poste, have pointed out (Tomlinson, 1989), in selecting proteins for therapeutic use, little rational thought has been applied to their site of action, namely, whether the putative therapeutic protein is to act systematically (i.e. endocrine-like), or whether it is a mimic of an endogenous molecule that is normally produced to act locally (i.e. as an autocrine- or paracrine-like mediator). Endogenous endocrine proteins (e.g. hormones such as insulin), act over long distances from their site of manufacture; they are also stable in blood and, if relevant, their size and surface character enable their (specific) extravasation. However, paracrine-/autocrine-like mediators are produced and released to act locally, and/or have very short chemical half-lives. Such properties ensure that they do not give rise to untoward effects on non-target neighbouring cells. Autocrine- and paracrine-like mediators are often produced at sites of inflammation, tumors and injuries (e.g. transforming growth factors alpha and beta, angiogenenin, fibroblast growth factors, and epidermal growth factor, etc.). As described elsewhere (Tomlinson, 1989), in the site-specific delivery of these mediators, one needs to consider the issues of chronicity in the activation of cells (including their temporal localisation and responsiveness) and, since such agents may be acting as part of a polymediator cascade of events, also the staging sequence through which they act.

CELL RECOGNITION AND PROCESSING

The pathway to a pharmacological site may involve passage into and through various cells. To survive and to maintain their function, both normal and diseased cells take up and process numerous types of materials by a variety of mechanisms. The uptake of hydrophilic macromolecules at the plasma membrane involves invagination and vesiculation of the lipid bilayer to form vesicles. These processes have a putative applicability for cell-selection and cell-access by site-specific therapeutic systems (Tomlinson, 1986). Movement along biological pathways can be either passive or active and include cell fusion, fluid-phase pinocytosis, phagocytosis, and both constitutive and non-constitutive receptor-mediated endocytosis. The binding to specific and/or non-specific regions of cell surfaces can aid other processes which cause cell access of macromolecules - including membrane fusion and simple diffusion. There are two classes of vesicular routings, i.e. those which involve constitutive recycling, and those which occur upon a specific ligand/receptor interaction. For the former class, these processes occur independently of external stimuli, and are for the purpose of imbibing, cell growth, and intra- and intercellular communication. Recognition and processing can be very sensitive processes, affected dramatically by slight changes in structure. For example, by analogy, an alteration in the gp 120 tryhptophan at position 432 of the HIV envelope can abrogate CD4 binding and thus affect its tropism (Cordonnier et al, 1989).

The capacity and kinetics of cell trafficking events are important considerations in the design of site-specific systems (as are the abundance, specificity, avidity and (cellular) fate of any receptor system). For the use of systems which rely on transport receptors to effect their site location, there is a need to know the dose/uptake relations. For example, studies on hEGF show an edxtraordinary dose-dependency of the pharmacokinetic behavior after iv administration in rats, due not to the saturation of excretion processes such as biliary and urinary excretions, but to the binding saturation of transport receptors (Murakami et al, 1989).

Site-specific delivery with soluble proteins relies on a combination of anatomical and (patho)physiological events, each bringing its own constraints and opportunities. These can involve either anatomically accessible and discrete compartments, as well as normal and dysfunctioning cellular processes of both a passive and an active type.

EXTRAVASATION OF MACROMOLECULES

To be effective, on occasions, macromolecular site-specific systems will need to leave the cardiovasculature in order to reach either extravascular-extracellular, and/or extravascular-intracellular target sites. Extravasation is under strict anatomical and (patho)physiological control. Hence, systems can either be incorporated into phagocytic cells which can extravasate, or pass directly through either interrupted endothelia, or through the cell barrier itself by exploiting fluid-phase and/or receptor-mediated, constitutive and non-constitutive cell transport processes.

Passage through Normal Endothelia via Passive Processes

The structure of the endothelium is complex and varies greatly in different organs and tissues. It is generally comprised of four layers, namely, the plasma membrane/plasma interface (which is formed by the glycocalyx of the cell and the proteins adsorbed onto it); the endothelium (a monolayer of cells which are metabolically very active and effect and monitor the bidirectional exchange of fluid between the plasma and the interstitial fluid); the basal lamina (which supports the endothelium); and the adventitia (a connective tissue which surrounds the lamina and fuses with the surrounding fibro-areolar tissue). Capillaries having a continuous endothelium and an uninterrupted basement membrane are the most widely distributed. Fenestrated capillaries are morphologically distinct from these, and are typified by having a very thin cytoplasm on each side of the nucleus (30-60 nm), and gaps of between 20-100 nm diameter at irregular intervals. Some tissues have sinusoidal endothelial membranes where the membrane is very thin and there is hardly any connective tissue separating the endothelial cells from the parenchymal cells of the underlying organ. These areas are often lined by phagocytic cells. Endothelial cells contain a large number of spherical vesicles of uniform diameter (plasmalemmelar vesicles). These are generally between 60 to 80 nm in diameter. Plasma molecules are selectively transported across the endothelium according primarily to their size, but their charge and their physicochemistry (i.e. hydrophilic/lipophilic balance) are also contributing factors. The capillary wall permeability for soluble macromolecules is well-documented. Soluble materials of less than 30 nm diameter are able to permeate through continuous endothelia (e.g. see Fig. 1). Rippe and Stelin (1989) have

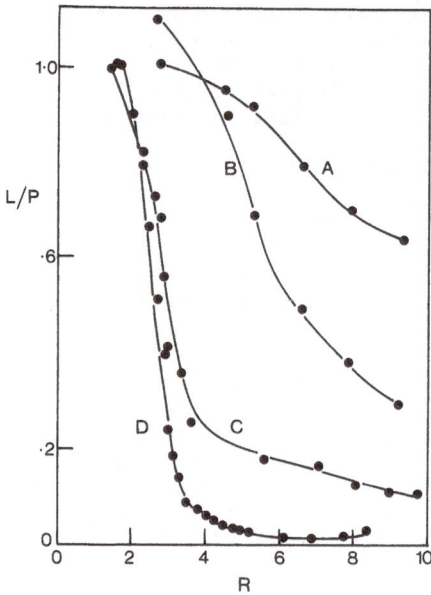

Fig. 1. Relationship between lymph-to-plasma ratios (L/P) for uncharged
dextrans and their molecular radius (R, nm) for different organs
in the rat; A-D are liver, intestine, leg and lung, respectively
(from Tomlinson, 1987). (Reproduced with the permission of the
copyright holder.)

recently examined the blood/peritoneal clearances of various endogenous
solutes in patients undergoing continuous ambulatory peritoneal dialysis.
They demonstrated that solute transport is compatible with a functional
blood/peritoneal barrier consisting of a two-pore membrane containing both a
large number of paracellular small pores of radius 4.0 to 5.5 nm, and a
smaller number of larger pores of radius 20 to 30 nm. In addition, they
found that whereas solutes smaller than 2.5 nm in radius permeated across
the peritoneal membrane mainly by diffusion across the small pores, solutes
larger than 4 nm were calculated to cross exclusively by unidirectional
convection across the large pores. Further, molecules larger than 2.5 to
3.0 nm radius (approximately 25 kDa) were simulated to be lost from the
peritoneal cavity by non-size-selective lymphatic drainage.

Extravasation of macromolecules occurs by diffusion and convection and
transcytosis through the vesicle/plasmalamellar pathways. For macromolecules, the large proportion of extravasation is due to convection. This is
related to the relative vascular and (interstitial) extravascular pressure,
and is porportional to the rate of fluid movement from the vessel lumen to
the interstitium. As pointed out by Jain (1989), this event is proportional
to the surface area and the difference between the vascular and the interstitial hydrostatic pressures minus the difference between the vascular and
interstitial osmotic pressures. Additionally, transmural arterial pressure
has an effect on endothelial (percolation-driven) transport of colloidal
particles (Fig. 2) (Chien et al, 1984). Since the osmotic reflection coefficient describes the effectiveness of the transluminal osmotic pressure
differences in causing movement of fluid across an endothelium, we can write
that the transport of a macromolecule across endothelia is characterised by
the three transport parameters of (i) vascular permeability, hydraulic
conductivity and reflection coefficient, (ii) surface area, and (iii) both
the transvascular concentration and pressure gradients (Jain, 1987).

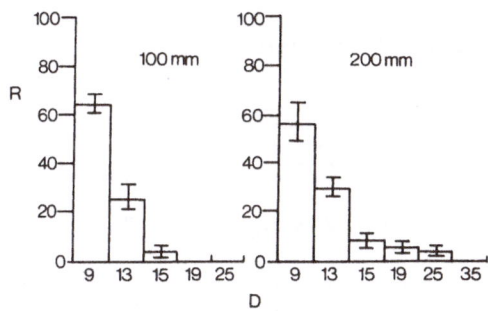

Fig. 2.   Effect of transmural pressure (100 mm Hg and 200 mm Hg) on the
distribution (R, percent) of Ag and Au colloid particles of
different diameter (D, nm) found in the carotid endothelium and
subendothelial space (in dogs) (Chien et al, 1984). (Reproduced
with the permission of the copyright holder.)

Pathophysiological Opportunities for Extravasation

Inflammation. The hyperpermeability of endothelial barriers at various
sites of inflammation is well established. This has been regarded as being
of potential use for the selective delivery of anti-inflammatory drugs to
inflamed extravascular regions. For example, in areas of inflammation
induced by carrageenin, the accumulation of lipid microspheres of approximately 200 nm diameter around endothelial cells of blood vessels, and the
penetration of these to the outer layer of blood vessels, has been reported.
Inflammation can cause regional changes in the structure, chemical composition and permeability of the endothelium. Permeability changes appear to be
due to the effect of histamine and bradykinin which act directly on the
capillary venule endothelia vessel wall, with various other mediators (including leukotriene B4 and the complement enzyme C5), effecting a rapid
interaction between venular endothelial cells and circulating neutrophils.
It is unclear what such hyperpermeability means in terms of the pathogenesis
of the underlying disease and the adequate retention of the carrier, particularly when one appreciates that inflamed sites often contain phagocytic
cells. Interestingly, intravenously administered radiolabelled 'small'
liposome particles can be used to image joints of patients affected by
rheumatoid disease (Williams et al, 1986); though it has been found that
when the disease is in remission, with no active synovitis, then accumulation of the radiolabel does not occur, suggesting that some accumulation
is due to phagocytic activity.

Ischemia/hypertensive vascular lesions. An increased permeability in the
endothelia is seen as an important factor in the pathogenesis of hypertensive lesions leading to infiltration and accumulation of plasma material.
For example, in experimental malignant hypertension, colloidal iron and
carbon particles of between 5 to 50 μm diameter are able to extravasate.
Capillaries have also been shown to be permeable at sites of tissue ischaem-
ia, both in the mesenteric artery and myocardium. However, this is not the
case with other hypertensive states.

Tumor endothelia. Ultrastructural studies of both animal and human extravascular tumors indicate that a significant fraction of tumor blood vessels
have wide interendothelial junctions, a large number of fenestrae and trans-
endothelial channels formed by vesicles, and discontinuous and/or absent
basal lamina. In addition, various tissue uptake studies have also demonstrated that the vascular permeability of tumors is significantly higher
than where there is a continuous membrane such as in skin and muscle.

Table 1. Average Time for MoAb and Fab to move in the tumour interstitium (from Jain, 1989)

Distance	Average time	
100 μm	about 0.5 h	about 0.2 h
1 mm	" 2–3 days	" 0.5–1 day
1 cm	" 7 months	" 2 months

However, macromolecular drugs such as immunotoxins and drug-antibody complexes, extravasate very poorly and percolate minimally into tumor masses. Jain (1989) has concluded that the poor extravasation of macromolecules could be due to tumors containing regions of high interstitial pressure, coupled with a decreased vascular pressure. This serves to lower fluid extravasation, which in turn leads to low levels of macromolecules extravasating since their transport normally occurs primarily by convention (Jain, 1987). (As tumors grow, their interstitial pressure increases correspondingly.) Also, since large tumors have a relatively lower vascular surface area than small vascularised tumors, then this should lead to a lower transvascular exchange in large tumors compared to small ones (Jain and Baxter, 1988). After extravasation, the flux of a macromolecule in the interstitium occurs via diffusion and convection, the former being proportional to the concentration gradient in the interstitium, and the latter to the interstitial fluid velocity (which, in turn, is proportional to the pressure gradient in the interstitium). In normal tissues, the matrix is composed of collagen, elastin, nidogen, etc. within which is the fluid and a hydrophilic gel made mainly of polysaccharides. In tumors, the large interstitial space and low concentrations of polysaccharides favor the movement of macromolecules in the tumor interstitium (Jain, 1989). But this event does not appear to occur readily, as indicated by a heterogeneous distribution in tumors of administered exogenous macromolecules. It appears that two events serve to inhibit such movement. First, and especially for systems targeted to bind to a surface marker, the binding of a macromolecule to an antigen will lower its apparent diffusion coefficient, and second, because of the large distances and the nature of the network, large molecules such as immunoglobulins can take considerable time to diffuse (Table 1). Related to this is the question of the fate of any low molecular weight drug released from a macromolecular carrier at the tumor surface. Notwithstanding that the concentration gradient for such a drug will permit its diffusion in all directions, Levin and coworkers have computed that the time taken for low molecular weight drugs to diffuse from a well-perfused tumor shell inwards to a poorly-perfused tumor core is long (Levin et al, 1980). For example, even for drugs with high diffusion coefficients (in the region of 10^{-6} to 10^{-5} cm^2 sec^{-1}), to travel 5 nm into the tumor mass would take between 24 and 48 hours. To achieve even this would require high levels of drug to be constantly present at the surface of the tumor mass. It is evident that this could lead to systemic toxicity.

There have been numerous attempts to selectively deliver drug to a tumor target and to maintain it there using various recognition ligands. These include hormones, porphyrins, lectins, sugars and anti-receptor monoclonal antibodies able to recognise some feature on the tumor surface (such as transferrin and low-density lipoprotein receptors). It appears that many of the markers suggested for targeting are not tumor-specific (except perhaps oncofetal markers), but that they are present in an abnormal abundance. In addition, it is known that tumor masses have a clonal heterogeneity – which may preclude the approach of using selective recognition

Table 2. Some Key (Patho)Physiological and Biochemical Issues Affecting
         the Pharmacokinetic and Pharmacodynamic Behaviour of Site-Specific
         Macromolecules Acting on Extravascular Tumours

Exclusivity, abundance and trafficking capacity of surface markers
Clonal heterogeneity and biochemical resistance
Intra/extravascular location
Diffusion, convection and binding of drug and/or drug/carrier complex
Multiple sites of drug resistance
Lower hydraulic conductance
A reducing relative endothelial surface
An increasing interstitial pressure
Lower blood pressure
Slow interstitial diffusion
Membrane permeability
Toxic side-effects
Non-linear dose/response relationship
Chronicity of growth and responsiveness

---

ligands in site-specific drug delivery due to the possibility for the selection of cells which have a biochemical drug resistance.

Thus, access to solid tumors is highly limiting for the use of macromolecular drugs/carriers in cancer, because (a) in humans, the dilation of tumor blood vessels leads to a lowering in hydraulic (transmural) conductance; (b) as tumor masses grow, their relative endothelial surface area (for extravasation) is reduced; (c) interstitial pressure increases, further reducing conductance; (d) the interstitial movement of macromolecules in tumor masses can be in the order of days to move a millimetre; and (e) specific (antigen) binding (by an antibody) serves to delay this movement and to reduce tumor cell penetration. Table 2 defines many of the factors which affect the pharmacokinetics and pharmacodynamics of cytotoxic macromolecular drugs. Clearly, a complex interplay exists between all of these parameters. The overwhelming conclusion that must be drawn from this is that macromolecular drugs are inappropriate agents for directly treating extravascular tumors located in tissues having a continuous endothelium. The same arguments do not apply to accessible tumors such as those in the spleen or the blood. Attempts to improve on the selectivity of anticancer agents has been made by an indirect two-stage approach, in which a bacterial enzyme carboxypeptidase G2 (CPG2) is conjugated to a F(ab')$_2$ fragment (against a subunit of human chorionic gonadotrophin). After localisation (which, following on from the above discussion, must be minimal), a prodrug of a cytotoxic agent able to be selectively cleaved by the enzyme, is introduced (Bagshawe et al, 1988). In mice this has been shown to lead to a reduction in tumor growth.

RETENTION IN THE CENTRAL COMPARTMENT

In many cases, macromolecular drugs and carriers need to persist in the central blood compartment. This may be either because of a need to access a target (cell) within the cardiovascular system, or to remain in the central compartment long enough to be able to remain intact and to extravasate (via either passive or active means). Size, surface charge, chemical stability, and surface physical and physicochemical stability, are the most important features for achieving this persistence. For therapeutic (site-specific) proteins needing to remain in the blood compartment, two prime methods have been adopted. Namely, increasing the apparent size of the protein and/or reducing its (untoward) interactions with blood and tissue components.

Uptake by the Mononuclear Phagocyte System (MPS)

Recognition of macromolecular therapeutic systems by the immune system is mediated through physicochemical interaction. Frequently, opsonization by fibrinogen, fibronectin and other blood components, is a prelude to recognition and thence removal by cells of the formed complex; antigen-antibody interaction and Fc-mediated removal also occurs. Opsonized materials are taken into cells by engulfment after adjerence to, and vesiculation of, phagocytosing cell membranes. Both opsonization and adherence can be diminished if the attractive forces between the interacting therapeutic protein, blood macromolecule and, for example, a cell-surface macromolecule are diminished. Adsorption and adhesion are complex phenomena which are controlled by many factors including hydration, and electrostatic, dispersion and steric forces, and by other short-range interactions (Norde, 1984). Interfacial adsorption is dependent upon a balance between these forces. Colloidal particles will attract each other through van der Waals interactions (short-range), and repel each other through long range, repulsive (e.g. Coulombic) forces. As proteins approach one another there is a net attraction, with a potential energy barrier to interaction at closer proximities, and with strong interaction at very short ranges. Interaction can be avoided by creating a high potential energy barrier. Although _in vitro_ this may be achieved by charge/charge effects, this is likely to be diminished _in vivo_. Napper and Netschey (1971) have argued that for (particulate) colloids, a high potential energy barrier can be formed by creating a sterically stabilized surface upon introducing a hydrated (i.e. hydrophilic) polymer at the surface of the colloid. The hydration effect is enthalpic in origin; the stabilization effect being manifested by both osmotic effects and chain entanglements, both of which are entropic in origin (Ottewill, 1977). The size of any repulsive barrier should be determined by both the thickness of the polymer layer and its density, as well as by polymer-polymer interactions caused by specific interactions along the polymer chain. It is probable that steric stabilization is akin to the mechanism whereby blood cells and various bacteria and parasites escape detection by the mononuclear phagocyte system (MPS). Surface modifications to proteins can be made to improve their tolerance within the vasculature, due largely to the formation of a surface which makes it energetically unfavorable for other macromolecules to approach.

Chemical Protectants

Many examples of hydrophilic (bio)polymeric protectants have been described for conjugation to therapeutic proteins. Synthetic and biological materials have been used or suggested, and include polyethylene glycols, poloxamers, poloxamines, albumin, immunoglobulin G, carboxymethycellulose, natural xanthans and sorbitans, etc. Conjugations of proteins with hydrophilic polymers have often been reported as being very successful in altering their potency as well as for reducing their immunogenicity and increasing their duration of action. Abuchowski and Davis (1977) first adopted this approach for stabilizing therapeutic proteins by forming protein conjugates with hydrophilic polyethylene glycol chains. Others have increasingly used this approach, and modifications of it, for lengthening the blood half-lives of a number of peptidergic mediators, including lymphokines, and enzymes, such as catalase, asparaginase and urokinase, whilst still maintaining their reactive functionalities. Steric stabilization (or polymer-excluded volume) approaches to avoiding opsonization have also received attention for modifying antibodies (Rihova et al, 1986), and the modification of antibodies with hydrophilic polymers can be additionally utilised to enable the chelation of small inorganics for dianosis purposes (Torchilin et al, 1986).

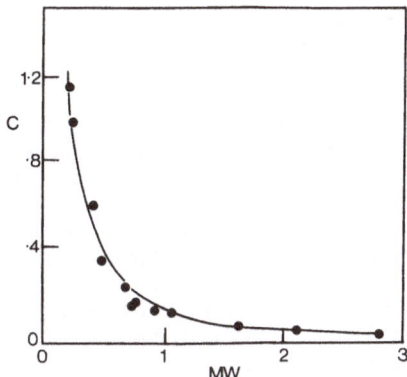

Fig. 3. Effect of size on the systemic clearance (C) of recombinant interleukin 2 (rIL-2) in rats, showing the relationship between C
$(ml.min^{-1})$ and effective molecular weight $(MW \times 10^{-5})$ of rIL-2
modified with polyethylene glycol (Katre et al, 1989). (Reproduced
with the permission of the copyright holder.)

Covalent attachment of a hydrophilic polymer such as PEG to a small
protein will significantly increase its size. This can be modulated by the
extent of protein modification and the size of polymer used. Figure 3 shows
the experimentally found relationship between the systemic clearance rates
of PEG-modified recombinant IL-2 in rats and their effective size and shows
that for such systems an abrupt change in clearance occurs at around 70kDa,
which of course is predictable from known information on serum proteins and
filtration (Katre et al, 1989). IL-2/PEG is reported to have a half-life 10
times that of non-pegylated IL-2, a lower toxicity, whilst still retaining
its antitumor activity in patients with advanced cancers.

White et al (1989) have shown that by prolonging the circulating time
of SOD and catalase with PEG, conjoint treatment with these increased survival and consistently decreased lung injury and neutrophil recruitment and
activation in rats exposed to hyperoxia. In addition, they found that poly-
ethylene-attached antioxidant enzymes decrease pulmonary oxygen toxicity (in
rats); indeed, in vitro studies even suggest that PEG itself may be contributing to protection by scavenging hydroxyl radical but not superoxide or
hydrogen peroxide.

A decrease in immunogenicity of therapeutic proteins may result from
either a reduction in their aggregation, or simply by a masking of any
antigenic determinants. It has been found that the primary and secondary
IgE antibody responses to protein may be suppressed by chemically conjugating protein with a derivative of polyaspartic acid, i.e. α,β-poly[(2-
hydroxyethyl)-DL-aspartamide) - chosen since it had been used as a plasma
expander without apparent toxicity (Okada et al, 1985). Although the
mechanism of this effect is not fully defined, it has been shown that the
suppressive effect of protein modified with polyethylene glycol (Lee et al,
1981) or fatty acid (Segawa, 1981) is due to the induction of suppressor T
cells and that conjugation with a copolymer of D-glutamate and D-lysine
leads to suppression via tolerance of B cells (Katz et al, 1972).

Conjugation of proteins with hydrophilic polymers can also increase
their chemical and physicochemical stability and is known to enhance resistance to proteolysis and heat denaturation (Wileman et al, 1986). Thus,
such chemical protection can affect clearance, and, perhaps, immune response.

CHANGING DISPERSION THROUGH STRUCTURAL MODIFICATION

Deletion Mutants

Site-directed mutagenesis can be used to create de novo heterologous proteins which may or may not broadly resemble endogenous material. These approaches are being used not only to improve on the stability and intrinsic specificity of such endogenous proteins, but also (and increasingly) to achieve a selective and often prolonged delivery of the polypeptide/protein to an active site. The efforts to develop further types of plasminogen activators serve as good examples. Recent work has pointed to the altered pharmacokinetic and thrombolytic properties of deletion mutations of human tissue-type plasminogen activator (tPA) in rabbits (Collen et al, 1988). Wild-type tPA is characterised by a rapid clearance by the liver, with an alpha distribution phase half-life of a few minutes. Using a series of deletion mutants (which included removal of fibronectin-like, epidermal growth factor type and glycosylation-site regions), this group have demonstreated that regions within tPA responsible for its liver clearance, its fibrin affinity and its fibrin specificity are not localised in the same structures. They argue that it appears possible to alter specific functions of tPA related to poor pharmacokinetics without decreasing its efficacy. Harber et al (1989) have recently reviewed how the tools of molecular biology and protein engineering may be used to develop 'safer and more effective' plasminogen activators. They describe both domain-deletion and site-directed mutagenesis techniques for the creation of new plasminogen activators, as well as chimaeric (or hybrid - see below) molecules. Harber et al (1989) have reviewed the use of domain deletions to produce a shortened form of single-chain urokinase-like plasminogen activator (scuPA), which does not have the NH_2-terminal kringle of scuPA. Whilst it appears that this shortened form of scuPA is not present in vivo, it is surprising to many observers that although it is only 14 aminoacids longer than the low molecular weight two-chain urokinase, low molecular weight scuPA has a similar selectivity for fibrin than is exhibited by scuPA, thus showing that the kringle is not needed for fibrin selectivity. The low molecular weight form also is resistant to plasminogen activator inhibitor I, which should assist in helping to increase its residence in the plasma.

Novel fibrinolytic enzymes have been made which retain the properties of the parent but have modified dispersion patterns. For example, Pohl et al (1989) have patented a fibrinolytically active plasminogen activator of the tissue type, in which both the growth factor domain and the Kringle 1 domain have been deleted, and where various point mutations have been made. These heterologous proteins exhibit different eliminations in vivo (Fig. 4).

In addition, since plasminogen activator inhibitor-1 (PAI-1) (a member of the serpin superfamily), serves to prevent systemic activation of plasminogen, groups have attempted to produce serpin-resistant variants of tPA using site-directed mutagenesis, which could serve to maintain the enzyme in an active form in the plasma (Madison et al, 1989).

Glycosylation

Many endogenous glycoproteins (i.e. serum glycoproteins, lysosomal enzymes and perhaps also sulfated pituitary glycoproteins such as chorionic gonadotropin) interact through their specific carbohydrate residues complexing with (oligosaccharide-specific) recognition systems on the plasma surfaces of target cells. Glycosylation patterns are thus signals used by the body to regulate the dispersion of its own glycoprroteins, as implicated for both enzyme and hormone disposition as well as immune surveillance, coagulation, etc.

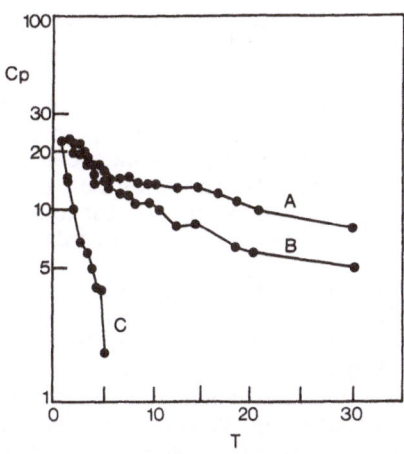

Fig. 4. Elimination of native and modified tissue plasminogen modifier
(tPA) from rabbit plasma (Cp is plasma concentration, mg.ml^{-1}) with
time (T, mins). A and B represent deletion mutants, and C is the
native full sized human tPA (Pohl et al, 1988). (Reproduced with
the permission of the copyright holder.)

     Three biological properties of glycoproteins may be adjusted upon
altering their surface distribution of carbohydrates, namely, i.e. circulating blood half-life (and potentially duration of action), immunogenicity, and ability to access (cellular) sites of action. Most glycoproteins contain three or more glycosylation sites, with some proteins
having 10 to 30 different sugars at each glycosylation site, with the
occupancy of each site varying. Expressed glycoproteins may be engineered
in order to affect their selectivity, and activity, for target sites.
Again, the extensive work being carried out to modify the glycosylation
patterns of tissue plasminogen activator, give that different levels and
types of glycosylation result in types of tPA differing in their ability to
cleave plasminogen after fibrin stimulation; and/or that a sugar positioned
at one of the Kringle rings could result in blockage of fibrin access.

     In contrast to possible changes in the amino-acid compositon of a pro-
tein, the numerous variations possible in linking simple sugars together
affords glycoproteins an almost limitless variability and diversity in
structure. Additional modifications, such as removal or addition of peripheral sugars and/or other functional groups such as acetyl, methyl, sulfate
and phosphate, are also possible. Oligosaccharides may be N-glycosidically
linked (to the peptide at Asn), or O-glycosidically linked (attached to Ser
and Thr). The oligosaccharides of the plasma glycoproteins are linked to
protein primarily through L-asparagine-N-acetyl-D-glucosamine. Recent
studies have examined the function of carbohydrate modifications of the
natural hematopoietins (including granulocyte-macrophage colony-stimulating
factor (GM-CSF), erythropoietin, and CSF-1). Extensive modifications produced by the addition of an asparagine-linked carbohydrate resulted in
rather heterogenous glycosylation patterns, dependent upon expression cell
and protein under study. However, it has been demonstrated that the effective half-life of a GM-CSF in the bloodstream of a rat is increased significantly by the addition of N-linked carbohydrate (Donahue et al, 1986).
Recent work has shown that by conjugating fragment A of diphtheria toxin to
a galactose-containing oligosaccharide such as asialofetuin or asialooroso-
mucoid, the resultant conjugate targets to hepatocytes, and is up to 3
orders of magnitude more toxic that the native fragment A. If galactosyl
residues are to be used clinically to target proteins to hepatocytes it is

important to appreciate that the in vivo uptake of galactosylated neoglyco-proteins has been shown to be highly dose-dependent (Vera et al, 1984).

Biotechnological processes. Glycosylation of a therapeutic protein may occur either during expression-cell processing, or post-expression via synthetic conjugation. Biotechnology processes use expression of the protein in a cell system. Glycosylation does not usually occur in pro-karyotic cells such as E. Coli. Some glycosylations do occur in eukaryotes, e.g. yeast and mammalian cells, however, the resultant glycosylation patterns can often be different, depending on the cells used. Also, al-though recombinant bacteria are able to produce large amounts of protein, these are often changed by bacterial proteases; within expression cells it is possible that expressed proteins of low solubility denature and form aggregates. Mammalian cells are particularly suitable both for expressing proteins with complex modifications such as γ-carboxylation of glutamyl residues, and for obtaining homologous (human) glycosylation patterns. Thus it is apparent that biotechnology processes can have a resultant marked effect on the biological disposition and efficacy of therapeutic proteins.

Protein re-modelling. Much recent progress has been made in the re-modelling of expressed glycoproteins intended for therapeutic use in order to affect their selectivity for target cells. Because of the complexity of these reactions, conventional carbohydrate chemistry is of little use and interest is centred on the modifications of proteins with oligosaccharides via enzymatic synthesis. Various chemical approaches to re-modelling proteins post-expression are being developed (e.g. Akiyama et al, 1987). Recently, an important new approach has been described which employs enzymes to elongate and terminate peripheral glycan chains of glycoproteins (Berger et al, 1986, 1987). Mammalian glycoproteins expressed in yeasts are likely to be substituted by mannans, and this group has been able to incorporate sialic acid into endo-β-N-acetylglucosaminidase H-treated oligomannose glycoproteins (Berger et al, 1986). The technology is of potential use in reducing any mannose-receptor uptake of these glycoproteins by cells of the MPS, etc. Although the approach needs optimizing in terms of enzyme activity, it demonstrates the successful incorporation of sialic acid into glycoproteins of the oligomannose type. Berger et al (1987) have also suggested the use of purified galactosyltransferase for the galactosylation of glycoproteins prior to their sialylation or their chemical linkage with oligosaccharides, and argue that this approach appears to be promising for the in vitro re-modelling of glycan chains in heterologous glycoproteins. For example, post-expression protein re-modelling has been used to resurface some of the enzymes indicated in lysosomal storage diseases. In attempts to improve on the affinity for glucocerebrosidase to macrophages, Furbish et al (1981) have reglycosylated the enzyme by simple conversion of the mannose-terminated core structures Man3GlcNAc$_2$ and Man3NAc(Fuc)GlcNAc by sequential treatment with neuraminidase, β-galactosidase and β-N-acetylhexosaminidase. This resulted in a five-fold more efficient increase in uptake by (rat) Kupffer cells over the native enzyme.

Candidate proteins reported amenable for this kind of re-modelling include tissue plasminogen activator, factor VIII, EPO, colony-stimulating factors, α_1-antitrypsin, β- and γ-interferons, interleukins I-III, and even antibodies, etc.

Deglycosylation. Conversely, protein deglycosylation should also affect dispersion (and stability and solubility) of a therapeutic glycoprotein. Indeed, deglycosylation (via chemical means using a mixture of sodium metaperiodate and sodium cyanoborohydride at low pH and temperature), has been shown to virtually eliminate the uptake of the toxin ricin A chain by the liver (Blakey and Thorpe, 1986). Ricin A chain is an oligomannosidic glycoprotein which has a very short half-life due to rapid removal by the mononuclear phagocyte system, and deglycosylation of the A chain may reduce removal by Kupffer cells. The chemical method described results in 'destruction' of approximately 50% of the mannose and most of the fucose residues, whilst leaving the N-acetylglucosamine and most of the xylose residues alone. Although chemical deglycosylation only marginally delays the clearance of deglycosylated ricin A from the blood compartment (in mice), this is probably due to the production of a smaller macromolecule able to be effectively filtered and excreted through the kidneys. Blakey and Thorpe (1986) assume that linking the deglycosylated ricin A chain to an immunoglobulin would diminish this later effect. Ricin A chain lacking carbohydrate side-chains has also been produced using recombinant technology (Vitetta et al, 1987).

Ligated Gene Fusion Hybrid Delivery Systems

Gene-fusion techniques may be used to produce distinct therapeutic proteins which combine the varied properties of parent proteins. This is exemplified by recent strategies which have been proposed for the targeting of bacterial and/or plant toxins to specific cells using hybrids created by ligating toxin and growth factor genes. Such an approach relies on the deletion of the toxin gene sequence encoding the cell-binding site, which allows the hybrid-fusion protein to display the cell specificity of the growth factor. Recently, a hybrid protein between interferon-γ and tumor necrosis factor-β has been shown to have a greatly increased antiprolif- erative activity in vitro, compared to either interferon-γ or TNF-β alone, whilst still retaining their antiviral activity and cytotoxic effects. Hybrid protein delivery systems may involve not just adduction of a protein (fragment) to a recognition moiety, but also to a re-ordering of the struc- ture of the efffector portions of therapeutic proteins in order to enhance their pharmacological action.

Harber et al (1989) have pointed to the use of hybrids as novel mole- cules that combine delivery and effector functions for use in plasminogen activator therapy. For example, high-affinity fibrin selectivity and resistance to plasminogen activator inhibitor I can be introduced into a molecule by construction of a hybrid composed of the A chain (fibrin binding domain) of tissue plasminogen activator, and the low molecular weight form of scuPA (i.e. the catalytic site of urokinase)in a form that is not suscep- tible to plasminogen activator inhibitor. Such enterprise however is not without its difficulties. As Haber et al (1989) point out, unfortunately the fibrin-binding activity of the recombinant hybrid is less than that of native tPA, with its fibrin selectivity being found in preliminary work to be less than that of single- or of two-chain tPA. Hybrid protein delivery systems may involve not just adduction of a protein (fragment) to a recog- nition moiety, but also to a re-ordering of the structure of the effector portions of therapeutic proteins in order to enhance their pharmacological action. For example, Murphy (1987) has described a hybrid protein, in sequence, and joined by peptide bonds, of the enzymatically active fragment A of diphtheria toxin, a fragment including the cleavage domain 1_1 adjacent to fragment A, a fragment which includes (at least) a portion of the hydro- phobic domain of fragment B (though not including the generalised eukaryotic binding site of fragment B), and finally a fragment which includes a portion of a cell-specific polypeptide able to bind the conjugate delivery system to a specific cell feature.

Recent work by Chaudray et al (1989), has produced a single chain antibody-toxin by the fusion of cDNAs encoding anti-Tac antibody variable regions with a fragment of DNA encoding a modified form of Pseudomonas exotoxin (anti-Tac is a monoclonal antibody to the p55 subunit of the human IL-2 receptor). Previous attempts at making such immunotoxins have suffered because the antigen binding site of the antibody is made from two separate

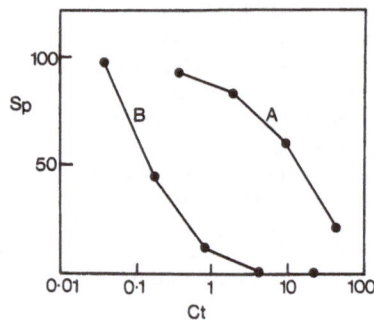

Fig. 5. Comparative _in vitro_ cytotoxicity of hybrid proteins, showing
covalently linked immunotoxin (A) and a fusion protein (B).
Cytotoxicity was measured as protein synthesis (Sp, percent of
control) at various concentrations of toxin (Ct, ng.ml^{-1})
(Chaudhary et al, 1989). (Reprinted (with modification) by permission from Nature, Vol. 339, pp. 394-397. Copyright (c) 1989,
Macmillan Magazines Ltd.)

polypeptide chains and this new approach has enabled a single chain antibody-toxin protein to be made in E. Coli which contains the two variable
domains of the antibody joined by peptide linkage and fused to a modified
exotoxin. Figure 5 gives some of their data showing the greater potency of
the fusion protein compared to the covalently linked protein, against cells
which strongly express the IL-2 receptor. The lower toxicity of the covalent hybrid protein is thought to be due to the linkage being through
lysine residues in domain III of the toxin, leading to a reduction in the
activity of the toxin. It is not proven that it is a general case that
fusion proteins are more active than covalently linked systems, but these
authors consider that their approach can be generally used to produce active
recombinant immunotoxins with (other) antibodies. In a similar fashion the
ricin A chain, which catalytically inhibits the 60S subunit of ribosomes _in
vitro_, linked to an anti-HIV gp120 monoclonal antibody, is considered to be
effective in the treatment of acute and chronic HIV infections (Pincus et
al, 1989).

Interesting further examples of this approach include the production of
phospholipid anchor domain (PAD) fused to a heterologous polypeptide, which
could for example be the anchor domain of mDAF (i.e. decay accelerating
factor). DAF conjugates or fusions serve to deliver DAF to target cells to
inhibit complement activation at the surfaces of such cells and and are
claimed to be useful in allograft rejection and autoimmune diseases (Caras,
1989).

Decoy Delivery Systems

Included in this category of novel proteins having both recognition and
effector functions, are the recently described immunoadhesins (Capon et al,
1989). Specifically, these are antibody-like molecules, containing the
gp120-binding domain of the receptor for human immunodeficiency virus. They
have been shown to block HIV-1 infection of T cells and monocytes.
Interestingly, such novel hybrids' delivery systems have a long plasma
half-life. The fusion of the gp120-binding domain of CD4 to the Fc domain
of an immunoglobulin has considerable merit when one appreciates that the
formed heterologous protein retains some of the important properties of both
of its parent molecules; namely, they bind gp120 and block infection of T
cells by lymphotropic HIV-1, and of monocytes by monocytotrophic HIV-1.
They are also comparable to antibodies in their long plasma half-life, as

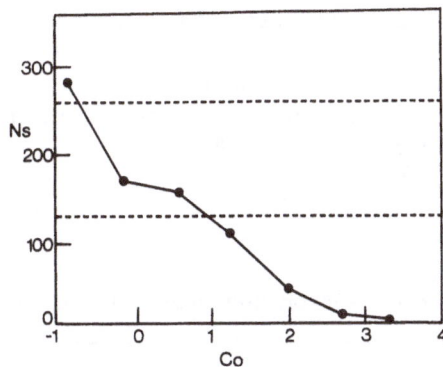

Fig. 6. Neutralisation of HIV with pentameric recombinant CD4-IgM fusion
proteins, showing the relationship between number of syncytia (Ns)
and concentration of recombinant protein (Co, $ng.ml^{-1}$). The dashed
lines indicate the number of syncytia in the absence of any recombinant protein and also the number at 50 percent inhibition
(Traunecker et al, 1989). (Reprinted (with modification) by
permission from Nature, Vol. 339, pp. 68-70. Copyright (c) 1989,
Macmillan Magazines Ltd.)

well as their ability to bind Fc receptors and protein A. The claims of
Capon et al (1989) include that the attainment of a high steady-state level
of immuno-adhesin makes it likely that effective concentrations of the
hybrid will be attained in lymph and lymphatic organs (where HIV may be most
active). Similarly, pentameric CD4-IgM chimeras have been produced
(Traunecker et al, 1989). These are shown to be 1000 times more active than
their dimeric CD4-IgG counterparts in syncytium inhibition assays (Fig. 6).
Also effector functions, such as the binding of Fc receptors and the first
component of the complement cascade(Clq) are retained (Traunecker et al,
1989), which is contrary to the results found with the CD4-IgG chimaeric
delivery decoys which contain the CH1 domain and do not bind Clq. This
suggests that it could be the deletion of the CH1 domain which leads to
complex aggregation.

Synthetically Linked Hybrid Conjugates

Large changes in the biological disposition of proteins have been
reported upon their chemical linkage to other protein (fragments). For
example, the toxin gelonin - which has a circulation half-life in mice
estimated at 3.5 minutes, when conjugated to immunoglobulin (fragments) has
a terminal phase blood half-life in the order of days, with only a slight
variation in this time as the conjugated immunoglobulin (fragment) is
changed (Scott et al, 1987). The immunotoxin field provides many relationships between protein structure and deposition. Point mutations in the B
polypeptide chain of diphtheria toxin that block non-specific binding to
non-target cells have been produced. Upon covalently linking this entity to
an anti-T-cell monoclonal antibody, it may be demonstrated that because of a
change in the non-target tissue distribution of the toxin, it becomes orders
of magnitude less toxic than the native toxin to non-target cells (in
vitro). The availability of monoclonal antibodies has led many workers to
consider these for the targeting of toxic materials to human tumors.

Taetle et al (1988) have recently prepared an immunoconjugate from an
antiepidermal growth factor (anti-EGF) receptor antibody and recombinant
ricin-A chain. Their data show that this site-specific conjugate gives a
dose-dependent killing of cells that express EGF receptors. However, and

Table 3. Some Factors Affecting the Pharmacodynamic Response

Rate and frequency of drug input
Rate of attainment of a critical plasma concentration
Presence of a drug-free interval
Infusion duration
Clearance
Frequency
Route
Dosing interval
Geometry of site of action and proximity of input to it

---

again typically for these types of anti-proliferative hybrids, the kinetics of cell killing with these conjugates was protracted, suggesting that prolonged exposure may be required for _in vivo_ anti-tumor effects.

A further interesting example is that due to Reiter and Fishelson (1989) who have covalently linked a monoclonal antibody to C3b. This conjugate may facilitate _in vitro_ and _in vivo_ tumor cell destruction by the host complement system. The complement system has the potential to destroy tumor cells but only if directed towards it. Reiter and Fishelson (1989) covalently linked the complement component C3b (which binds to the surface of complement activators) to anti-transferrin receptor antibodies, in the belief that this receptor is expressed to a greater extent on rapidly proliferating cells than on non-proliferating cells. Whilst this approach is imaginative, the previous arguments given above relating to the access and persistence of macromolecular constructs at extravascular solid tumors apply.

RELATION BETWEEN INPUT AND DRUG PHARMACODYNAMICS

Although it is largely outside the scope of this contribution, it must also be recognised that there are a number of features of drug administration or input that can be identified which affect the drug's pharmacodynamic properties. These are given in Table 3 and are discussed more fully elsewhere (Tomlinson, 1989, 1990a,b). It is apparent that the constant drug blood levels are often an inappropriate feature for administration of therapeutic protein systems.

FORMULATION

The successful development of macromolecular drugs and drug carriers is a challenge due to their size, complexity, conformational requirements and their often complicated analytical, stability and solubility profiles (Tomlinson, 1987; Pearlman and Nguyen, 1989). Although again outside the scope of this contribution, it is clear (Pearlman and Nguyen, 1989) that one of the greatest challenges for developing such products will be the successful preclinical formulation of such molecules which will require relevant and discriminatory analytical methods.

CONCLUDING REMARKS

This brief contribution has pointed to some of the challenges and developments in the use of proteinaceous therapeutic systems for clinical use. Unfortunately, a belief still persists that macromolecular drugs can be used for successfully treating tumors residing in tissues having a continuous endothelium. The arguments and examples presented in this paper are intended to show how this has a minimal chance for success.

However, the tremendous advances being made using the tools of modern molecular cell biology are enabling new types of site-specific therapeutic drug to be fashioned. These will be able to take advantage of unique features of accessible lesions in order to effect treatment for a variety of important diseases.

REFERENCES

Abuchowski, A., Van Es., T., Palczuk, N.C. and Davis, F.F., 1977, Alteration of immunological properties of bovine serum albumin by covalent attachment of polyethylene glycol, J.Biol.Chem., 252:3578.

Akiyama, A., Bednarski, M., Kim, M-J., Simon, E.S., Waldmann, H. and Whitesides, G.M., 1987, Enzymes in organic synthesis, Chem.Brit., 645.

Bagshawe, K.D., Springer, C.J., Searle, F., Antoniw, P., Sharma, S.K., Melton, R.G. and Sherwood, R.F., 1988, A cytotoxic agent can be selectively generated at cancer sites, Br.J.Cancer, 58:700.

Berger, E.G., Greber, U.F. and Mosbach, K., 1986, Galactosyltransferase-dependent sialylation of complex and endo-N-acetylglucosaminidase H-treated core N-glycans in vitro, FEBS Lett. 203:64.

Berger, E.G., Muller, U., Aegerter, E. and Strous, G.J., 1987, Biology of galactosyltransferase: recent developments, Biol.Chem.Trans., 15:610.

Blakey, D.C. and Thorpe, P.E., 1986, Effect of chemical deglycosylation on the in vivo fate of ricin A-chain, Cancer Drug Delivery, 3:189.

Capon, D.J., Chamow, S.M., Mordenti, J., Marsters, S.A., Gregory, T., Mitsuya, H., Byrn, R.A., Lucas, C., Wurm, F.M., Groopman, J.E., Broder, S. and Smith, D.H., 1989, Designing CD4 immunoadhesins for AIDS therapy, Nature, 337:525.

Caras, I.W., 1989, Nucleic acid and methods for the synthesis of novel fusion polypeptides with a phospholipid anchor, Intnl. Patent Appl., Publ.9.2.89, WO 89/01041.

Chaudhary, V.K., Queen, C., Junghans, R.P. and Walkmann, T.A., Fitzgerald, D.J. and Pastan, I., 1989, A recombinant immunotoxin consisting of two antibody variable domains fused to Pseudomonas exotoxin, Nature, 339:394.

Chien, S., Fan, F., Lee, M.M.L. and Handley, D.A., 1984, Effects of arterial pressure on endothelial transport of macromolecules, Biorheology, 21:631.

Collen, D., Stassen, J-M. and Larsen, G., 1988, Pharmacokinetics and thrombolytic properties of deletion mutants of human tissue-type plasmingen activator in rabbits, Blood, 71:216.

Cordonnier, A., Montagnier, L. and Emerman, M., 1989, Single amino-acid changes in HIV envelope affect viral tropism and receptor binding, Nature, 340:571.

Donahue, R.E., Wang, E.A., Kaufman, R.J., Foutch, L., Leary, A.C., Witek-Giannetti, J.S., Metzger, M., Hewick, R.M., Steinbrink, D.R., Shaw, G., Kamen, R. and Clark, S.C., 1986, Effects of N-linked carbohydrate on the in vivo properties of human GM-CSF, Cold Spring Harbor Symposia on Quantitative Biology, Vol. LI:685.

Furbish, F.S., Steer, C.J., Krett, N.L. and Barranger, J.A., 1981, Uptake and distribution of placental glucocerebrosidase in rat hepatic cells and effects of sequential deglycosylation, Biochim.Biophys.Acta, 673:425.

Harber, E., Quertermous, T., Matsueda, G.R. and Runge, M.S., 1989, Innovative approaches to plasminogen activator therapy, Science, 243:51.

Jain, R.K., 1987, Transport of molecules across tumor vasculature, Cancer Metastasis Rev., 6:559.

Jain, R.K. and Baxter, L.T., 1989, Delivery of novel therapeutic agents in tumors: physiological barriers and strategies, J.Natl.Cancer Inst., 81:570.

Jain, R.K. and Baxter, L.T., 1988, Mechanisms of heterogeneous distribution of monoclonal antibodies and other molecules in tumors: significance of elevated interstitial pressure, Cancer Res., 48:7022.

Katre, N.V., Young, J.D. and Zimmerman, R., 1989, Chemical modification of interleukin 2 with polymers: a potent drug-delivery system, in: "Therapeutic Peptides and Proteins", Current Communications in Molecular Biology, Cold Spring Harbor Laboratory.

Katz, D.H., Hamaoka, T., and Benecerraf, B., 1972, Immunological tolerance in bone marrow-derived lymphocytes. I. Evidence for an intracellular mechanism of inactivation of hapten-specific precursors of antibody-forming cells, J.Exp.Med., 136:1410.

Lee, W.Y., Sehon, A.H. and Akerblom, E., 1981, Suppression of reaginic antibodies with modified allergens. IV. Induction of suppressor T cells by conjugates with polyethylene glycol (PEG) and monomethoxy PEG with ovalbumin, Int.Archs.Allergy Appl.Immunol., 64:100.

Levin, P.A., Patlak, C.C. and Landhal, H.D., 1980, Heuristic modelling of drug delivery to malignant brain tumours, J.Pharmacokinetics Biopharm., 8:257.

Madison, M.L., Goldsmith, E.J., Gerard, R.D., Gething, M-J.H. and Sambrook, J.F., 1989, Serpin-resistant mutants of human tissue-type plasminogen activator, Nature, 339:721.

Murakami, T., Kishimoto, M., Higashi, Y., Amagase, H., Fuwa, T. and Yata, N., 1989, Down-regulation and its effect on epidermal growth factor receptors on the pharmacokinetics of human epidermal growth factor after i.v. administration in rats, Int.J.Pharmaceutics, 54:259.

Murphy, J.R., 1987, Hybrid Protein, U.S. Patent, 4,675,382, June 23, 1987.

Norde, W., 1984, Physicochemical aspects of the behaviour of biological components at solid liquid interfaces, in: "Microspheres and Drug Therapy. Pharmaceutical, Immunological and Medical Aspects", S.S. Davis, L. Illum, J.G. McVie and E. Tomlinson, eds., Elsevier, Amsterdam.

Napper, D.H. and Netschey, A., 1971, Studies of the steric stabilisation of colloidal particles, J.Colloid Interface Sci., 37:528.

Okada, M., Matsushima, A., Katsuhata, A.,Aoyama, T., Ando, T. and Inada, Y., 1985, Suppression of IgE antibody response against ovalbumin by the chemical conjugate of ovalbumin with a polyaspartic acid derivative, Int.Archs.Allergy Applied.Immunol., 76:79.

Ottewill, R.H., 1977, Stability and instability in disperse systems, J.Colloid Interface Sci.,58:357.

Pearlman, R. and Nguyen, T.H., 1989, Formulation strategies for recombinant proteins: human growth hormone and tissue plasminogen activator, in: "Therapeutic Peptides and Proteins", Current Communications in Molecular Biology, Cold Spring Harbor Laboratory.

Pohl, G., Hansson, L. and Lowenadler, B., Novel fibrinolytic enzymes, Europ.Patent Appl., 14.06.88, Publ. No. 0 297 A1.

Pincus, S.H., Wehrly, K., Cheseboro, B., 1989, Treatment of HIV tissue culture infection with monoclonal antibody-ricin A chain conjugates, J.Immunol., 142:3070.

Reiter, Y. and Fishelson, Z.V.I., 1989, Targeting of complement to tumor cells by heteroconjugates comprised of antibodies and of the complement component C3b, J.Immunol., 142:2771.

Rihova, B., Kopecek, J., Kopeckova-Rejmanova, P., Strohalm, J., Plocova, D. and Semoradova, H., 1986, Bioaffinity therapy with antibodies and and drugs bound to soluble synthetic polymers, J.Chromatogr., 376:221.

Rippe, B. and Stelin, G., 1989, Simulations of peritoneal solute transport during CAPD. Application of two-pore formalism, <u>Kidney Int.</u>, 35:1234.

Scott, C.F., Lambert, J.M., Goldmacher, V.S., Blattler, W.A., Sobel, R., Schlossman, S.F. and Benecerraf, B., 1987, The pharmacokinetics and toxicity of murine monoclonal antibodies and of gelonin conjugates of these antibodies, <u>Int.J.Immunopharmac.</u>, 9:211.

Segawa, A., Borges, M.S., Yokota, Y., Matsushima, A., Inada, Y. and Tada, T., 1981, Suppression of IgE antibody response by the fatty acid-modified antigen, <u>Int.Archs.Allergy Appl.Immunol.</u>, 66:189.

Taetle, R., Honeysett, J.M. and Houston, L.L., 1988, Effects of anti-epidermal growth factor (EGF) receptor antibodies and an anti-EGF receptor recombinant-ricin A chain immunoconjugate of growth of human cells., <u>J.Natl.Cancer Inst.</u>, 80:1053.

Tomlinson, E., 1986, (Patho)physiology and the temporal and spatial aspects of drug delivery, <u>in</u>: "Site-Specific Drug Delivery", E. Tomlinson and S.S. Davis, eds., John Wiley & Sons, Chichester.

Tomlinson, E., 1987, Theory and practice of site-specific drug delivery, <u>Advanced Drug Delivery Reviews</u>, 1:87.

Tomlinson, E., 1989, Considerations in the physiological delivery of thera-peutic proteins, <u>in</u>: "Novel Drug Delivery and its Therapeutic Applic-ations", L.F. Prescott and W.S. Nimmo, eds., John Wiley & Sons Ltd., Chichester.

Tomlinson, E., 1990a, Control of the biological dispersion of therapeutic proteins, <u>in</u>: "Protein Design and the Development of New Therapeutics and Vaccines", G. Poste and S.T. Crooke, eds., Plenum Press, New York, in press.

Tomlinson, E., 1990b, Selective delivery of protein drugs, <u>in</u>: "Protein Production, T.J.R. Harris and C.C.G. Hentschel, eds, in press.

Torchilin, V.P., Khaw, B.A., Klibanov, A.L., Slinkin, M.A., Haber, E. and Smirnov, V.N., 1986, Modification of monoclonal antibodies by poly-mers possessing chelating properties, <u>Bull.Exp.Biol.Med.</u>, 102:946.

Traunecker, A., Schneider, J., Kiefer, H. and Karjalainen, K., 1989, Highly efficient neutralisation of HIV with recombinant CD4-immunoglobulin molecules, <u>Nature</u>, 339:68.

Vera, D.R., Krohn, K.A., Stadalnik, R.C. and Scheibe, P.O., 1984, Tc-99m-galactosyl-neoglycoalbumin: <u>in vivo</u> characterisation of receptor-mediated binding to hepatocytes, <u>Radiology</u>, 151:191.

Vitetta, E.S., Fulton, R.J., May, R.D. and Uhr, J.W., 1987, Redesigning nature's poisons to create anti-tumor reagents, <u>Science</u>, 238:1098.

White, C.W., Jackson, J.H., Abuchowski, A., Kazo, G.M., Mimmack, R.F., Berger, E.M., Freeman, B.A., McCord, J.M. and Repine, J.E., 1989, Polyethylene glycol-attached antioxidant enzymes decrease pulmonary oxygen toxicity in rats, <u>J.Appl.Physiol.</u>, 66:584.

Wileman, T.E., Foster, R.L. and Elliott, P.N.C., 1986, Soluble asparaginase-dextran conjugates show increased circulatory persistence and lowered antigen reactivity, <u>J.Pharm.Pharmacol.</u>, 38:264.

Williams, B.D., O'Sullivan, M.M., Saggu, G.S., Williams, K.E., Williams, L.A. and Morgan, J.R., 1986, Imaging of rheumatoid arthritis using liposomes labelled with technetium, <u>Brit.Med.J.</u>, 293:1143.

RECOMBINANT LIGAND-TOXIN CONJUGATES, DOMAIN ENGINEERING AND THE SEARCH FOR

TARGETED PHARMACEUTICALS

Marco Soria

Biotechnological Research, Farmitalia Carlo Erba, 24 Viale
E. Bezzi, Milano and Institute of Pharmacological Sciences
University of Milano, Italy

INTRODUCTION

Affinity therapy is the opportunity to attempt precise targeting of
therapeutic effectors to their site of action. Within the past years,
advances in biochemistry and molecular biology of membranes and cell growth
regulation have provided tools that can assist in the design of more selective chemotherapeutic agents. Cloning and expressing genes coding for a
variety of effector molecules allows the assembly of recombinant chimaeras
for anticancer, immunosuppressive, antiviral and cardiovascular targeted
therapy (Soria and Zeller, 1978; Soria and Martini, 1987; Soria, 1989a).
Site-specific delivery might thus fulfil Paul Ehrlich's dream of "magic
bullets" by engineering specific structural/functional domains into therapeutic agents, in order to improve their persistence in the circulation and
to localize them at the site of action.

PROTEIN DOMAINS

That protein domains, or at least coherent structural motifs, could be
transferred between unrelated genes to obtain new proteins out of "old"
domains, rests on the theory of "exon shuffling", in which exons are described as units in the evolutionary rearrangements of genes. Many large
eukaryotic proteins have been shown to consist of characteristic domain
structures, often reflected in the genome by the arrangement of the exons
and introns making up the gene (Gilbert, 1978).

The mosaic character of many proteins presenting independently folding
domains (Patthy, 1985) has now been confirmed in several protein families,
thus revealing fascinating insights in evolutionary cut-and-pasting.
Examples of these are to be found in the coagulation/fibrinolytic pathway,
with the "kringle" domains found in plasminogen, in prothrombin, in tissue-
type plasminogen activator (tPA), in pro-urokinase (single-chain urokinase-
type plasminogen activator, scU-PA) and in apolipoprotein(a). All of them
must contribute to the dramatic interplay between thrombosis and atherogenesis (Scott, 1989). tPA and scU-PA are very similar in their COOH-
terminal domains, where the catalytic activity of both enzymes is located.
Fibrin binding of tPA has been found to reside in the amino terminal segment
of tPA, with four domains involved. The amino terminal domain of scU-PA has
now been shown to interact with a specific receptor on normal and neoplastic
cells (Blasi, 1988). The complex interplay between scU-PA receptor-mediated

Targeting of Drugs, Edited by G. Gregoriadis *et al.*
Plenum Press, New York, 1990

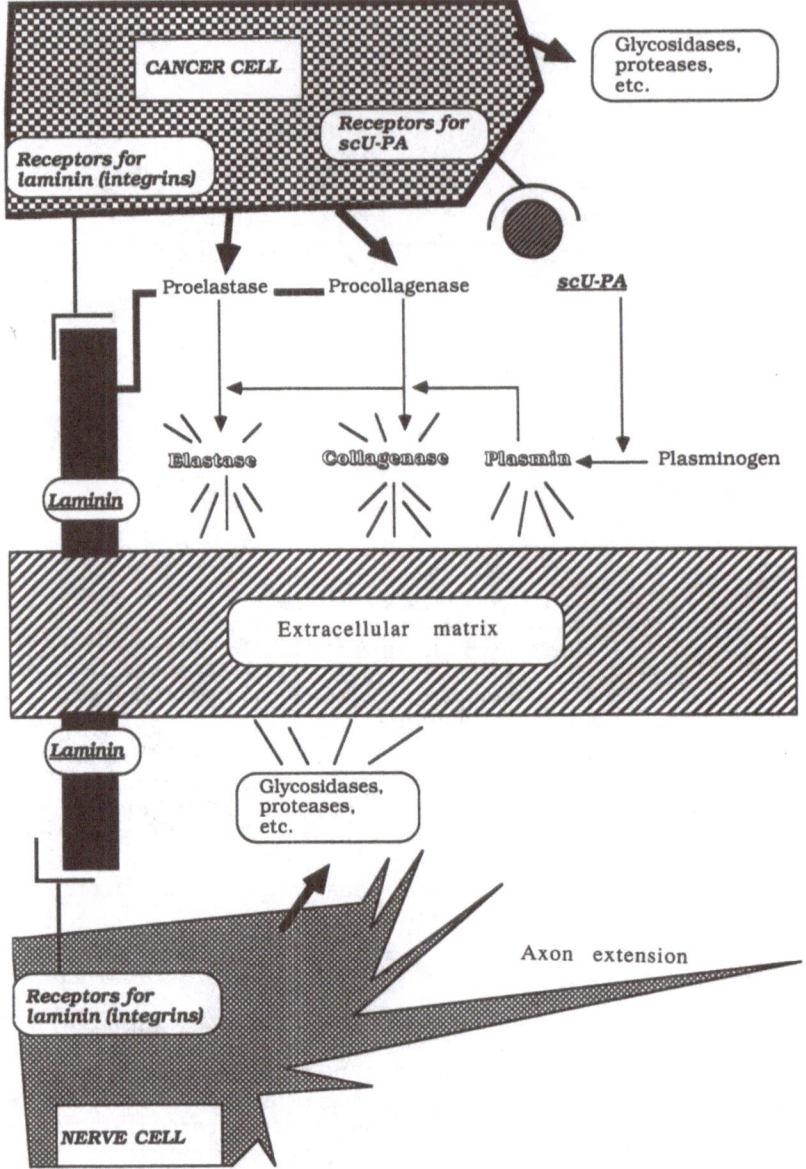

Fig. 1.   Parallelism between some embryonic neuronal interactions (i.e. growth cone extension) and metastatic spreading of cancer cells.

fibrinolysis and other proteolytic-glycolytic activities at the surface of tumor cell metastases is schematically illustrated in Fig. 1, with striking parallelisms emerging from molecular studies on embryonic neuronal inter- actions (Dodd and Jessel, 1988).

Examples of domain engineering performed by mother Nature were recently provided at least in two instances: a newly identified chimeric glucose 6- phosphate dehydrogenase (G6-PD), whose amino terminal segment is encoded by a different gene residing on a separate chromosome (Kanno et al, 1989), and in the discontinuous genetic system of ribosomal rRNA of Clamidomonas reinhardtii, where large transcripts of scrambled RNA have been found inter- spersed with protein coding sequences (Boer and Gray, 1988).

Table 1. Mitotoxins

(a) Semi-recombinant:

Recombinant bFGF conjugated biochemically to SO-6	(Lappi et al, 1989)

(b) Totally recombinant

    b1. Diphtheria Toxin +

αMSH	(Murphy et al, 1986)
IL2	(Williams et al, 1987; Kelley et al, 1988)

    b2. Pseudomonas Exotoxin +

IL2	(Lorberboum-Galski et al, 1988; Bailon et al, 1988; Case et al, 1989; Lorberboum-Galski et al, 1989)
TGFα	(Chaudhary et al, 1987)
IL4	(Ogata et al, 1989)
IL6	(Siegall et al, 1988)
soluble CD4	(Chaudhary et al, 1988)
anti-Tac (Fv)	(Chaudhary et al, 1989)

TARGETING BY DOMAIN ENGINEERING

Chimaeric cytotoxins

Recombinant DNA methodologies have been applied to obtaining targeting protein chimaeras, especially in the anti-cancer, immunosuppressive and antiviral fields. Cloned fragments of diphtheria toxin and of Pseudomonas exotoxin A have been fused to genes coding for growth factor ligands (such as IL-2, IL-4, IL-6 or TGF-α), soluble receptors for viral proteins such as CD4, or Fv immunoglobulin fragments of anti-Tac antibodies (Table 1).

Plant toxin genes have not yet been used to construct a totally recombinant mitotoxin, although successful attempts at expressing a fused gene coding for staphylococcal protein A and a functional ricin A chain in E. coli have been reported (Kim and Weaver, 1988). Of the single-chain (Type I) RIPs, SO-6 (also called saporin-6), obtained from the seeds and leaves of Saponaria officinalis, has been recently cloned in our laboratory (Benatti et al, 1989; Soria, 1989b). Progress with plant RIPs as partners for genetic conjugates has been slower than their bacterial counterparts. However, E. coli expression of a synthetic gene coding for a Mirabilis jalapa RIP is an important step in this direction (Habuka et al, 1989).

Conversely, a mitotoxin obtained by biochemical conjugation of "natural" SO-6 to recombinant human basic Fibroblast Growth Factor (bFGF) has been obtained and shown to effectively kill bFGF receptor-bearing cells.

Preliminary attempts at expressing recombinant SO-6 in E. coli have been performed by fusing the coding region of the gene to that of β-galacto-sidase (provided by plasmid pUEX 3; Bressan and Stanley, 1987). The result-ing chimeric protein of the expected molecular weight is recognized by immunoblot analysis with a polyclonal anti-SO6 antibody (Lorenzetti et al, 1988).

Adhesive and Topogenic Domains

Engineering of adhesive and topogenic domains in chimaeric proteins either by recombinant techniques or by attaching effector peptides has been extensively reviewed previously, including examples of practical appli-cations (Soria, 1989a). In a recent report, the RGD motif of fibronectin and other adhesive proteins has been employed as a peptide conjugate to bovine serum albumin in a model system based on CHO attachment to extra-cellular matrix components. While RGD-albumin constructs adsorbed to plastics promoted cell adhesion, soluble constructs behaved as inhibitors of integrin-mediated cell adhesion (Danilov and Juliano, 1989). Since bacter-ial expression systems for albumin are available (Latta et al, 1987), it should not be too long before constructs of this type will be obtained by recombinant techniques.

Recent advances in the elucidation of intracellular traffic have led to the identification of various "topogenic sequences" that direct the pro-teins' fate intracellularly. Besides signal peptides, a variety of signals and motifs confer to specific protein domains their intracellular locali-zation or migration. Short signal sequences have been identified as nuclear location sequences (NLS), and synthetic peptides containing NLS are able to cause nuclear accumulation when cross-linked to albumin or to immunoglobulin G and microinjected in Xenopus oocytes (Goldfarb et al, 1986; Lanford et al, 1986) or in the cytoplasm of different cell lines (Yoneda et al, 1988; Newmeyer and Forbes, 1988). An extension of this approach has allowed nuclear targeting, both in vitro and in vivo, of plasmid DNA that was trans-ferred into cells by means of vesicle complexes. A procedure was developed to form vesicle complexes consisting of ganglioside liposomes entrapping DNA mixed with Sendai virus, fused to red blood cell membrane ghosts entrapping a nuclear DNA-binding protein. Thus fusogenic Sendai virus on the surface of the vesicles could mediate intracellular delivery of the nuclear protein, that in turn directed plasmid DNA to the nucleus. The DNA was rapidly transported in the nuclei of cultured cells, and was shown to be efficiently expressed in the nuclei of liver cells after injection in the portal vein of rats (Kaneda et al, 1989a, b). In the future, it might be possible to sub-stitute the entire Sendai virus particles with fusogenic components derived from viral proteins. Synthetic peptides derived from influenza virus hema-gglutinin (HA) or from vesicular stomatitis virus spike glycoprotein G have been used to mimic viral membrane destabilization and fusion (Schlegel and Wade, 1984; Stegmann et al, 1989), and might thus be engineered in an appropriate fashion to effect targeted liposomal fusion or to confer fuso-genic properties to other proteins.

Hormone Receptors

The hormone-binding domains of various steroid hormone receptors, in-cluding the receptors for thyroid hormones and retinoic acid, have been characterized by cDNA cloning of these multi-domain members of a regulatory superfamily (Gronemeyer, 1988). When expressed in eukaryotic cells, the cloned hormone-binding domain confers to receptor-negative cells the ability to bind steroid hormones with the same affinity and specificity as the corresponding receptor. Fusion proteins containing the hormone-binding domains of the chicken progesterone receptor have been obtained fused to the carboxy-terminus of β-galactosidase, and possess the same binding character-istics as their "natural" counterparts (Eul et al, 1989). These results provide very useful tools for therapeutic and diagnostic applications: since progestin binding can be monitored even in intact bacterial cells, this might be useful for screening steroidal derivatives with agonistic or antagonistic activities using permeabilized cells.

Other recombinant chimaeras with hormone-binding domains fused to reporter protein-coding genes have been reported recently: the human 2-adrenergic receptor was fused to β-galactosidase and was shown to retain its pharmacological properties (Marullo et al, 1988). The binding domains for 1 and for 2 adrenergic receptors were then fused to a fragment of an E. coli gene that is expressed at the bacterial cell surface. In this manner, a ten-fold increase in active receptor number was achieved while simplifying the experimental procedures for measuring binding of several ligands. Their affinity and selectivity for the chimaeric receptors was similar to those observed in mammalian tissues (Marullo et al, 1989).

Ferritin is a heteropolymer consisting of two homologous subunits, H and L, that assemble to form a hollow shell composed of 24 subunits. Each H chain monomer has a size of 183 amino acids and its cDNA has been success-fully expressed to high levels in E. coli (Levi et al, 1987). The remark-able resistance to protease treatment and heat denaturation of this protein lends itself to useful exploitation of the unusual characteristics of the ferritin core for therapeutic and diagnostic applications (Luzzago and Cesareni, 1989). It is easily foreseen that in the future many other types of similar tools will be constructed and employed in an expanding variety of procedures, in fields that will not be limited to medicine but will cover all sorts of yet unanticipated applications.

REFERENCES

Bailon, P., Weber, D.V., Gately, M., Smart, J.E., Lorberboum-Galski, H., Fitzgerald, D., and Pastan, I., 1988, Purification and partial characterization of an Interleukin 2-Pseudomonas exotoxin fusion protein, Biotechnology 6:1326.
Benatti, L., Saccardo, B., Dani, M., Nitti, G., Sassano, M., Lorenzetti, R., Lappi, D.A. and Soria, M., 1989, Molecular cloning and character-ization of the gene coding for a ribosome inactivating protein from the leaves of Saponaria officinalis, Eur.J.Biochem., 183:465.
Blasi, F., 1988, Surface receptors for urokinase plasminogen activators, Fibrinolysis, 2:73.
Boer, P.H. and Gray, M.W., 1988, Scrambled ribosomal RNA gene pieces in Chlamidomonas reinhardtii mitochondrial DNA, Cell, 55:399.
Bressan, G.M. and Stanley, K.K., 1987, pUEX, a bacterial expression vector related to pEX with univeral host specificity, Nucl.Acids Res., 15:10056.
Case, J.P., Lorberboum-Galski, H., Lafyatis, R., Fitzgerald, D., Wilder, R.L. and Pastan, I., 1989, Chimeric cytotoxin IL2-PE40 delays and mitigates adjuvant-induced arthritis in rats, Proc.Natl.Acad.Sci. USA, 86:287.
Chaudhary, V.K., Fitzgerald, D.J., Adhya, S. and Pastan, I., 1987, Activity of a recombinant fusion protein between transforming growth factor type alpha and Pseudomonas exotoxin, Proc.Natl.Acad.Sci.USA, 85:4538.
Chaudhary, V.K., Mizukami, T., Fuerst, T.R., Fitzgerald, D.J., Moss, B., Pastan, I. and Berger, E.A., 1988, Selective killing of HIV-infected

cells by recombinant human CD4-Pseudomonas exotoxin hybrid protein, Nature, 335:369.

Chaudhary, V.K., Queen, C., Junghans, R.P., Waldmann, T.A., Fitzgerald, D.J. and Pastan, I., 1989, A recombinant immunotoxin consisting of two antibody variable domains fused to Pseudomonas exotoxin, Nature, 339:394.

Danilov, Y.N. and Juliano, R.L., 1989, (Arg-Gly-Asp)n-albumin conjugates as a model substratum for integrin-mediated cell adhesion, Exp. Cell Res., 182:186.

Dodd, J. and Jessell, T.M., 1988, Axon guidance and the patterning of neuronal projections in vertebrates, Science, 242:692.

Eul, J., Meyer, M.E., Tora, L., Bocquel, M.T., Quirin-Stricker, C., Chambon, P. and Gronemeyer, H., 1989, Expression of active hormone and DNA-binding domains of the chicken progesterone receptor in E.Coli, EMBO J., 8:83.

Gilbert, W., 1978, Why genes in pieces?, Nature, 271:501.

Goldfarb, D.S., Gariepy, J., Schoolnik, G. and Kornberg, R.D., 1986, Synthetic peptides as nuclear localization signals, Nature, 322:641.

Gronemeyer, H., 1988, Affinity labelling and cloning of steroid and thyroid hormone receptors: techniques, application and functional analysis, Ellis Horwood, New York.

Habuka, N., Murakami, Y., Noma, M., Kudo, T. and Horikoshi, K., 1989, Amino acid sequence of Mirabilis antiviral protein, total synthesis of its gene and expression in E. coli, J.Biol.Chem., 264:6629.

Kaneda, Y., Iwai, K. and Uchida, T., 1989a, Introduction and expression of the human insulin gene in adult rat liver, J.Biol.Chem., in press.

Kaneda, Y., Iwai, K. and Uchida, T., 1989b, Increased expression of DNA co-introduced with nuclear protein in adult rat liver, Science, 243:375.

Kanno, H., Huang, I-Y, Kan, Y.W. and Yoshida, A., 1989, Two structural genes on different chromosomes are required for encoding the major subunit of human red cell glucose 6-phosphate dehydrogenase, Cell, 58:595.

Kelley, V.E., Bacha, P., Pankewycz, O., Nichols, J.C., Murphy, J.R. and Strom, T.B., 1988, Interleukin 2-diphtheria toxin fusion protein can abolish cell-mediated immunity in vivo, Proc.Nat.Acad.Sci.USA, 85:3980.

Kim, J. and Weaver, R.F., 1988, Construction of a recombinant expression plasmid encoding a staphylococcal protein A-ricin A fusion protein, Gene, 68:315.

Lanford, R.E., Kanda, P. and Kennedy, R.C., 1986, Induction of nuclear transport with a synthetic peptide homologous to the SV40 T antigen transport signal, Cell, 46:575.

Latta, M., Knapp, M., Sarmientos, P., Brefort, G., Becquart, J., Guerrier, L., Jung, G. and Mayaux, J.F., 1987, Synthesis and purification of mature serum albumin from E. coli, Biotechnol. 5:1309.

Levi, S., Cesareni, G., Arosio, P., Lorenzetti, R., Soria, M., Sollazzo, M., Albertini, A. and Cortese, R., 1987, Characterization of human ferri-tin H chain synthesized in E. coli, Gene, 51:269.

Lorberboum-Galski, H., FitzGerald, D., Chaudhary, V.K., Adhya, S. and Pastan, I., 1988, Cytotoxic activity of an interleukin 2-Pseudomonas exotoxin chimeric protein produced in Escherichia coli, Proc.Nat. Acad.Sci.USA, 85:1922.

Lorberboum-Galski, H., Barrett, L.V., Kirkman, R.L., Ogata, M., Willingham, M.C., Fitzgerald, D.J. and Pastan, I., 1989, Cardiac allograft in mice treated with IL2-PE40, Proc.Natl.Acad.Sci.USA, 86:1008.

Lorenzetti, R., Benatti, L., Dani, M., Lappi, D.A., Saccardo, B.M. and Soria, M., 1988, Nucleotide sequence encoding plant ribosome-inactivating protein, Brit. Pat. Appl. n. 8801877.

Luzzago, A. and Cesareni, G., 1989, Isolation of point mutations that affect the folding of the H chain of human ferritin in E. coli, EMBO J., 8:569.

Marullo, S., Delavier-Klutchko, C., Eshdat, Y., Strosberg, A.D. and Emorine, L.J., 1988, Human beta2-adrenergic receptors expressed in E. coli retain their pharmacological properties, Proc.Natl.Acad.Sci.USA 85:7551.

Marullo, S., Delavier-Klutchko, C., Guillet, J-G., Charbit, A., Strosberg, A.D. and Emorine, L.J., 1989, Expression of human beta 1 and beta 2 adrenergic receptors in E. coli as a new tool for ligand screening, Biotechnol., 7:923.

Murphy, R.J., Bishai, W., Borowski, M., Miyanohara, A., Boyd, J. and Nagle, S., 1986, Genetic construction, expression, and melanoma-selective cytotoxicity of a diphtheria toxin-related gamma-melanocyte-stimulating hormone fusion protein, Proc.Natl.Acad.Sci.USA, 83:8258.

Newmeyer, D.D. and Forbes, D.J., 1988, Nuclear import can be separated into distinct steps in vitro: nuclear pore binding and translocation, Cell, 52:641.

Ogata, M., Chaudhary, V.K., Fitzgerald, D.K. and Pastan, I., 1989, Cytotoxic activity of a recombinant fusion protein between interleukin 4 and Pseudomonas exotoxin, Proc.Natl.Acad.Sci.USA, 86:4215.

Patthy, L., 1985, Evolution of the proteases of blood coagulation and fibrinolysis by assembly from modules, Cell, 41:657.

Schlegel, R. and Wade, M., 1984, A synthetic peptide corresponding to the NH2 terminus of vesicular stomatitis virus glycoprotein is a pH-dependent hemolysin, J.Biol.Chem., 259:4691.

Scott, J., 1989, Thrombogenesis linked to atherogenesis at last? Nature, 341:22.

Siegall, C.B., Chaudhary, V.K., Fitzgerald, D.J. and Pastan, I., 1988, Cytotoxic activity of an interleukin 6-Pseudomonas exotoxin fusion protein on human myeloma cells, Proc.Natl.Acad.Sci.USA, 85:9738.

Soria, M., 1989a, Molecular targeting and delivery: applications of recombinant DNA technology, Biotechnol.Appl.Biochem., in press.

Soria, M., 1989b, Immunotoxins, ligand-toxin conjugates and molecular targeting, Pharmacol. Res., in press.

Soria, M. and Martini, D., 1987, Recombinant strategies in the search for targeted pharmaceuticals, in: Proc. 4th European Congress on Biotechnology, Vol. 4, O.M. Neijssel, R.R. Van der Meer and K.Ch.A.M. Luyben, eds., Elsevier, Amsterdam.

Soria, M. and Zeller, L., 1978, Expectations and priorities of pharmaceutical industry, in: "Genetic Engineering: Scientific Discoveries and Practical Applications", H. Boyer and S. Nicosia, eds., Elsevier, Amsterdam.

Stegmann, T., Doms, R.W. and Helenius, A., 1989, Protein-mediated membrane fusion, Ann.Rev.Biophys.Chem., 18:187.

Williams, D.P., Parker, K., Bacha, P., Bishai, W., Borowski, M., Genbauffe, F., Strom, T.B. and Murphy, J.R., 1987, Diphtheria toxin receptor binding domain substitution with interleukin-2: genetic construction and properties of a diphtheria toxin-related interleukin-2 fusion protein, Prot.Eng., 1:493.

Yoneda, Y., Imamoto-Sonobe, N., Matsuoka, Y., Iwamoto, R., Kiho, Y. and Uchida, T., 1988, Antibodies to Asp-Asp-Glu-Asp can inhibit transport of nuclear proteins into the nucleus, Science, 242:275.

EXTRAVASCULAR DIFFUSION AND CONVECTION OF ANTIBODIES IN TUMORS

Leon Cobb

Division of Experimental Pathology and Therapeutics
Medical Research Council Radiobiology Unit, Harwell
Didcot, Oxon. OX11 ORD, UK

INTRODUCTION

With the advent of hybridoma technology came the ability to prepare
large volumes of monoclonal antibody (McAb). (In this presentation the McAb
referred to is usually of immunoglobulin G. For work with antibody
fragments see references: Buchegger et al (1986), Buraggi et al (1985),
Carrasquillo et al, 1984), Covell et al (1986), Harwood et al (1985), Mach
et al (1980), Mach et al (1983), Wahl et al (1983)). Antibody could be prepared which was specific for antigens that were at a higher concentration in
tumors than in other tissues, so called tumor-associated antigens (TAAs).
Only rarely are McAbs cytotoxic in their own right and to have an effect on
tumors it is usually necessary to attach a cytotoxin. Cytotoxins may need
to be delivered either to the surface of each and every tumor cell (drugs,
plant toxins) (Baldwin, 1985; Thorpe, 1985) or simply into the general tumor
area (radionuclides, enzymes for activating pro-drugs) (Epenetos et al,
1987; Humm and Cobb, 1990; Bagshawe, 1988; Cobb and Humm, 1986). It is
clearly important to know the likely intratumor distribution of any injected antibody down to the microscopic level if we wish to assess whether the
cytotoxin-carrying antibody is likely to be effective. Our work on this
problem has been predominantly carried out on slow-growing animal tumors,
using I-125-labelled McAbs to TAAs. The animals are sacrificed at intervals
after McAb injection and the histological distribution of McAb in relation
to the target cell studied using auto-radiography. In an attempt to "bridge
the gap" between the experimental animal tumor data and the situation in
patients we have also studied the distribution of the host's own antibody
using frozen sections of human tumors together with immunoperoxidase staining for immunoglobulin.

It is important to stress the very wide variations in tumor histology
and vascular architecture that exist in both human and animal tumors. While
one tumor type may have an extensive, highly permeable, vasculature another
may be virtually avascular with resulting minimal egress of antibody. On
occasion such differences in vascularity may be seem within one tumor in a
single microscopic field. This means that one part of a tumor in a patient
or animal may receive adequate quantities of cytotoxin-bearing antibody
while an adjacent area is starved of antibody. It is the aim of this
presentation to indicate the main factors which can influence intratumor
antibody distribution and at the same time to indicate very generally the

Targeting of Drugs, Edited by G. Gregoriadis *et al.*
Plenum Press, New York, 1990

29

types of tumors which would seem to be suitable for antibody-targeted
therapy.

VASCULAR ARCHITECTURE

Normal Tissue

In man and animals antibody can gain access to at least a part of the
surface of cells (e.g. the basal surface of epithelial cells) within minutes of intravenous injection (Taylor and Granger, 1984). The main except-
ions to this rapid access are cartilage and bone where intercellular matrix
is dense, organs such as brain where there is a specialized vascular barrier
and some cells of stratified epithelia such as skin. Extravasation and subsequent re-entry of antibody takes place predominantly through and between
the endothelial cells of the capillaries and post-capillary venules
(Karnovsky, 1968; Bennet et al, 1959). Any antibody not reabsorbed into the
blood stream is cleared by the thin-walled low pressure channels of the
lymphatic drainage system (Keele et al, 1982). Although a tumor can stimulate the host to produce new capillaries to support the expanding mass of
cells it would appear to be unable to stimulate the formation of new lymphatic channels. This failure to form adequate drainage channels can lead to
a patchy accumulation of tissue fluid and antibody throughout the tumor.

Capillaries vary greatly in permeability to immunoglobulin; from the
low permeability vessels of skin and lung, through the moderately permeable,

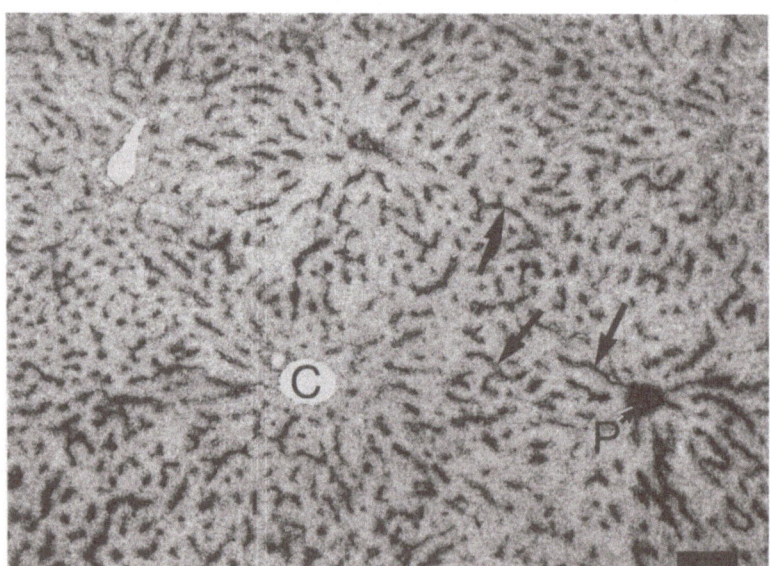

Fig. 1. Autoradiograph of normal mouse liver. 160 μCi of ^{125}I-labelled
non-specific McAb (80 ug) were injected intravenously and 6 h later
the liver fixed in formalin and subsequently processed. The sinusoids of the liver plus the perivascular space (arrows) contain the
labelled antibody and appear black in the figure. The disintegration of I-125 has caused grain exposure in the overlying
emulsion. Between the black sinusoids are single columns of pale
grey hepatocytes. Some part of the surface of each hepatocyte will
be adjacent to a sinusoid. P, portal area; C, central vein.
Stain, hematoxylin and eosin. The bar represents 80 μm.

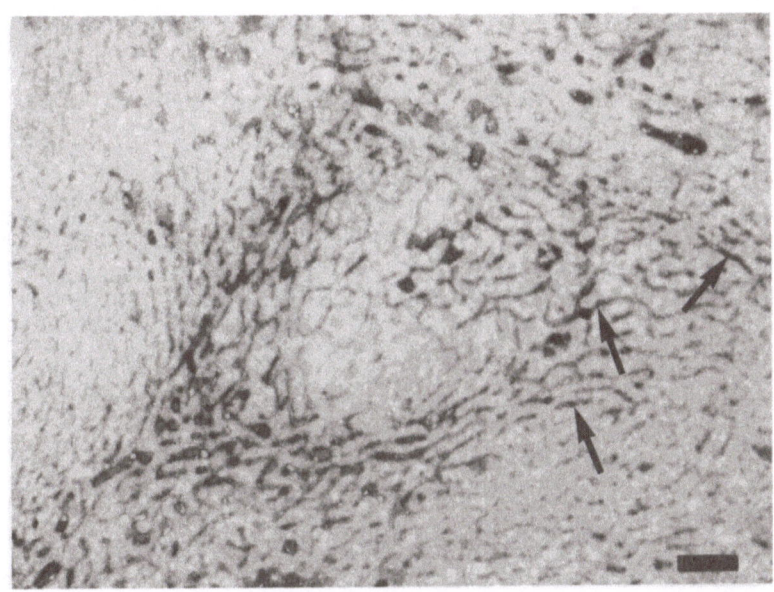

Fig. 2.   Autoradiograph of well differentiated hepatocellular carcinoma in a
          mouse. I-125-labelled non-specific McAb (IgG) had been injected 6
          h previously. The sinusoids and perivascular space (arrows) separ-
          ate columns of malignant hepatocytes. It is unlikely that most of
          the tumor cells are immediately adjacent to a blood vessel and
          therefore injected antibody would rapidly gain access to target
          antigens on the cell surface. Stain, hematoxylin and eosin. The
          bar represents 80 μm.

fenestrated, vessels of endocrine glands, gut and liver, to the highly permeable discontinuous endothelium of spleen and bone marrow (Bennet et al,
1959; Gray, 1980). As a very general statement it can be said that tumor
vessels are somewhat more permeable than those of the tissue from which the
tumor was derived (Dvorak et al, 1988). In addition there is some evidence
that tumors can dictate the structure of the supporting capillaries. For
example, thyroid gland has a fenestrated capillary network and the observation of Ludatscher and her colleagues (1979) is that carcinoma of the
thyroid has similarly fenestrated capillaries. If this is found to be a
general phenomenon tumors arising from cells normally supplied with permeable capillaries would be potentially more accessible to injected antibody
than those arising from a tissue with a continuous endothelial lining to the
blood vessels (Hirano and Zimmerman, 1972; Schelin, 1962; Wang and Campiche,
1982).

Carcinoma

     Well differentiated tumor of epithelial and glandular origin retains a
histological appearance similar to the tissue of origin (parent tissue). A
well differentiated carcinoma of the colon or rectum continues to form
tubular acini surrounded by a capillary network, and the well differentiated
follicular carcinoma of the thyroid continues to form distinct colloid-
filled follicles invested with an extensive capillary network. Well differentiated carcinoma would therefore seem to be a good candidate for antibody-
targeted therapy as extravasated McAb will quickly contact target cells
(Figs. 1-3). An exception arises with some tumors of the large intestine,
breast and lung (for example) where the host, for reasons that are not
understood, produces an extensive fibroblastic reaction (desmoplastic
reaction) between the epithelial elements, which can greatly increase the

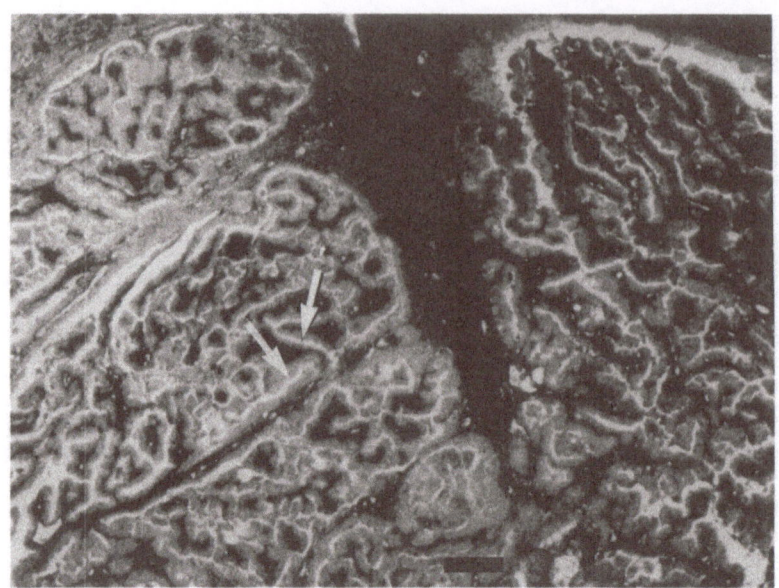

Fig. 3. Autoradiograph of well differentiated type II pneumonocyte tumor of
mouse lung. The black areas represent grains in the emulsion overlying the section. The grains are produced by the disintegration
of the I-125 conjugated to a non-specific IgG McAb injected intravenously 24 h before death. McAb in the vascular and perivascular
space has easy access to the tumor cells. In this figure the tumor
cells appear as rows of cells (grey) with the base of each cell
abutting the perivascular space (black). The apices of the cells
are bathed in a surfactant-like secretion which appears in the
figure as a white line (arrows). Stain, hematoxylin and eosin.
Bar represents 300 μm.

distance between groups of target cells and the blood vessels, and may itself hinder free movement of antibody (Fig. 4).

The common progression of tumor is for clones of less well differentiated cells to emerge (de-differentiation). The cells in these clones
usually have a shorter cell cycle time and therefore eventually predominate
numerically. De-differentiated cells tend to change from orderly growth in
sheets or acini to large nests of cells surrounded by stroma which carries
the vasculature (Fig. 5). The result of this is that the cells toward the
centre of the nests have reduced access to nutrients and to targeting antibody. Should the tumor cells retain their ability to bind firmly one to the
other by desmosomes, tight junctions, adhesive glycoproteins, etc. antibody
will be excluded from all but the outer layer of such nests (Sunderland et
al, 1987). The emergence of such clones in a tumor would reduce its suitability as a candidate for antibody-targeted therapy.

Sarcoma

The well differentiated sarcoma, by definition, resembles closely its
tissue of origin. Many of the non-skeletal sarcomas, e.g. muscle and

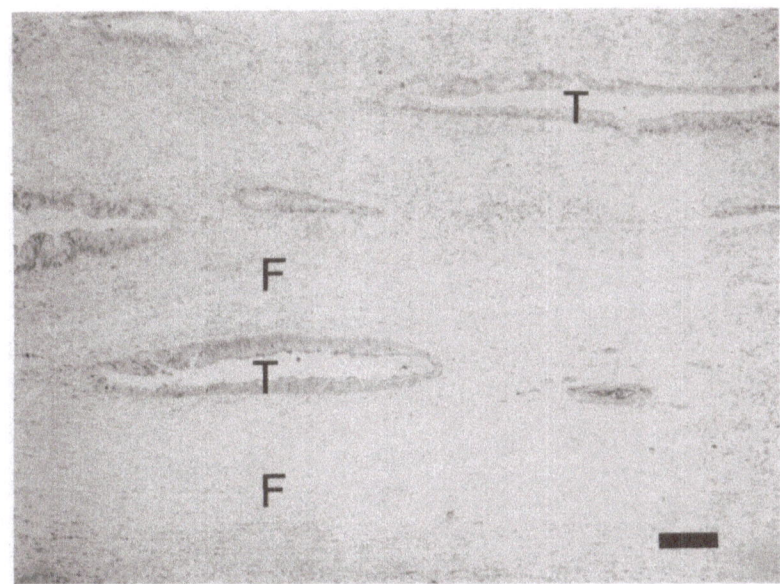

Fig. 4. Carcinoma of colon (man). The tumor cells are growing in the form
of tubules closely resembling the glandular crypts of normal colon.
The tubules (T) are separated by extensive fibroblastic reaction
(F) filling about 90% of the photomicrograph field. The dense
collagen with associated proteoglycans will impede the diffusion
and convection of injected McAb. Stain, hematoxylin and eosin.
Bar represents 200 μm.

lymphohemopoeitic tissue, have an adequate capillary network and reasonable
access for antibody. A problem arises with tumors of skeletal origin
(osteosarcoma and chondrosarcoma) as there is frequently extensive production by the tumor cells of extracellular matrix which may be highly impermeable to antibody (Jain, 1987a; Nugent and Jain, 1984a; Iozzo, 1987; Sweet et
al, 1979). De-differentiation commonly brings with it a loss of ability to
form intercellular matrix and other things being equal the cells of the de-
differentiated sarcoma should be more accessible to McAb than the well
differentiated counterpart.

INTRATUMOR PRESSURE

Examination of vascular architecture and tumor cell patterns provides
much information on the likely distribution of antibody to the tumor cell
population. However, the extravasation and retention of antibody within the
extracellular fluid is a very complex process governed by numerous factors.
The extravasated antibody moves through the tissue fluid by the processes of
diffusion and convection (Swabb et al, 1974). Diffusion is the result of
random movement of the antibody molecules and is gradient related, while
convection is movement of the antibody due to the mass flow of the interstitial fluid and is pressure related. These two processes have not yet
been separately measured within tumors but there is much current research
into this problem (Jain, 1987a, 1989; Nugent and Jain, 1984b).

Macromolecules accumulated in the extracellular fluid directly from the
capillaries, together with the continuous breakdown of cells within tumors,

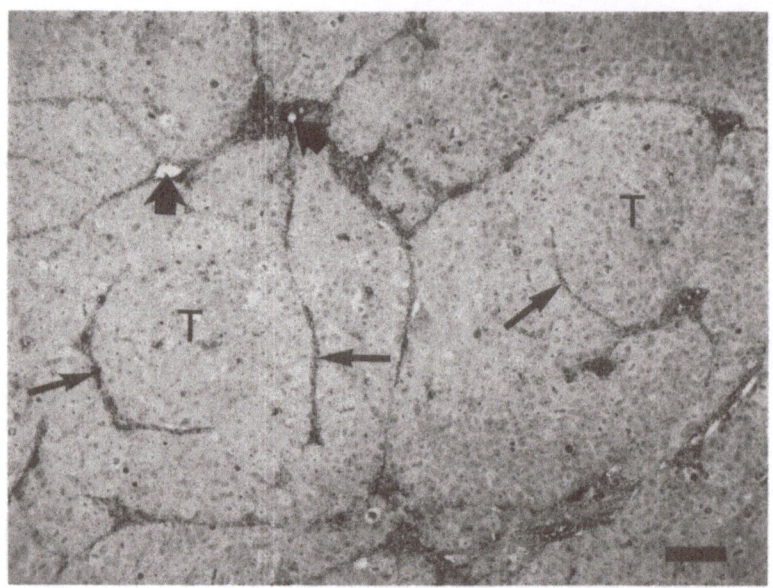

Fig. 5. Autoradiograph of poorly differentiated mouse mammary adenocarcinoma. I-125-labelled non-specific McAb had been injected
intravenously 24 h previously. The antibody remains restricted to
the stroma (arrows) surrounding the nests of tumor cells (T). The
blood vessels (arrow heads) are carried in the sheets of stroma.
Ultrastructural examination showed tight interdigitation of the
tumor cells and numerous desmosomes. Such close apposition of the
cells would reduce the access of antibody to the tumor cells, particularly those towards the centre of the nests. Stain, hematoxylin and eosin. Bar represents 80 μm.

cause an increase in extravascular osmotic pressure, and this in turn produces an increase in extravascular fluid retention. This increase in extravascular mass together with the increase in cell mass due to proliferation
causes an increase in intratumor pressure (ITP) (Hori et al, 1986; Wiig et
al, 1982; Clauss and Jain, 1988). As the extravascular pressure approaches
the intravascular pressure the blood flow inevitably decreases and with it
the supply of antibody (Jain, 1987). In time the center of the tumor, which
usually has the highest ITP, ceases to have any blood flow and develops
necrosis (Paskins-Hurlburt et al, 1982). The above is a somewhat simplified
description of a very complex process but the net result is that with the
growth of a tumor the amount of retained antibody per unit mass becomes
gradually reduced.

ANTIGEN DISTRIBUTION AND ANTIBODY/ANTIGEN INTERACTION

The distribution of antigen as assessed by immunohistological methods
has been reviewed elsewhere (Edwards, 1985). Briefly, the distribution of
TAAs can vary greatly throughout a tumor. An antigen present in high
density on differentiated tumor cells may be absent from the surface of
cells in a clone of de-differentiated cells. In other circumstances the
target antigen may be present in high density on resting cells but absent
from cells in active proliferation. Even when the antigen distribution is
uniform and the movement of macromolecules has theoretically little interference, antibody distribution can be grossly irregular. We have studied

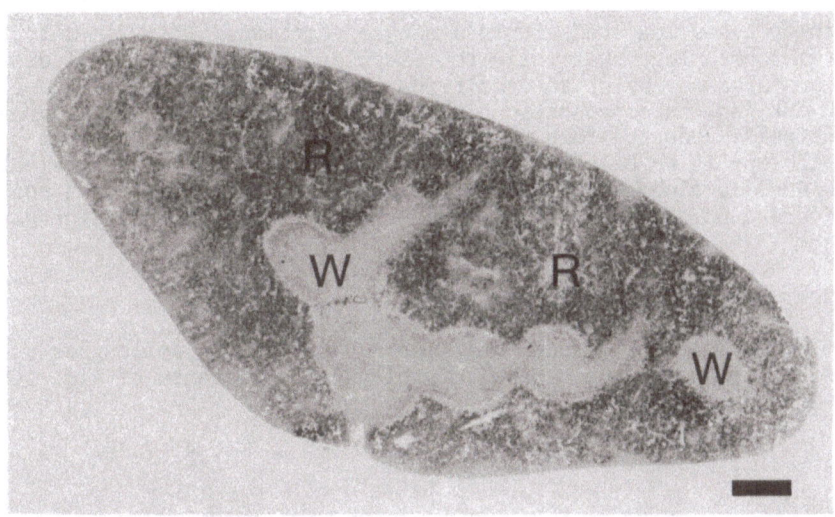

Fig. 6. Autoradiograph of lymphoma-infiltrated mouse spleen. The tumor is
a T-cell lymphoblastic lymphoma. In this specimen the tumor cells
are about 70% of the cell mass of the red pulp (R) and about 95% of
the cell mass of the white pulp (W). A tumor-specific antibody
(I-125-labelled) was injected intravenously and the mouse killed
after 6 h. Over 95% of the grains were found above the red pulp
and the remainder above the white pulp. The failure to penetrate
the tumor-infiltrated white pulp was thought to be due to retention
of the McAb close to the vessels of egress in the red pulp. In
another animal non-specific antibody distributed uniformly throughout the spleen within 30 min of injection. Stain, hematoxylin and
eosin. Bar represents 1 mm.

this situation in an experimental lymphoma (Cobb et al, 1987; Cobb et al,
1990). Mice with advanced splenic lymphoma were injected intravenously with
one of two ^{125}I-labelled McAbs. One antibody had a high affinity for a
surface antigen on the lymphoma cells and the other was a non-specific antibody not recognising the lymphoma cells. The spleens were examined by auto-
radiography 30 min and 6 h after injection of one or other of the antibodies. By 30 min the non-specific McAb was uniformly distributed throughout the tumor-infiltrated spleen and this indicated that there was no
mechanical hindrance to antibody movement. At both 30 min and 6 h the
specific antibody remained in the area surrounding the permeable sinusoids
of the spleen - which are in the red pulp (Fig. 6). The antibody was
apparently bound to lymphoma cells in the red pulp. The white pulp which
contained a high concentration of lymphoma cells (about 95%) showed little
or no antibody accumulation. Our present interpretation of this situation
is that the high affinity of the specific antibody for the antigen caused it
to be retained in the area of the cells first encountered after extravasation. The volume of specific McAb used was equivalent to 160 mg in man.
The solution to this problem might be to use an antibody with a lower affinity or to inject a significantly greater volume of antibody (where that is
practicable). Alternatively, repeated treatment might produce a gradual
erosion of the tumor beginning at the pericapillary zone.

CONCLUSIONS

Our work on antibody distribution in human and animal tumors indicates
that there can be a large variation in the pattern of antibody distribution

between tumor types and even within a single specimen. It is an unfortunate fact of tumor pathology that while the vascular structure and cell distribution in one area may point to conditions for extensive interaction of antibody with antigen, these conditions may not prevail in an adjacent area, or in a metastasis. Despite this very real problem we suggest if antibody targeted therapy is to be effective it is most likely to be against well differentiated carcinomas and poorly differentiated sarcomas - and possibly anaplastic carcinomas and sarcomas where the cells are loosely aggregated.

Acknowledgements

I am pleased to acknowledge help and advice from my colleagues S.A. Butler and J. Nolan and the careful histological assistance of A.J. Austin and H.L. Gordon.

REFERENCES

Bagshawe, K.D., 1988, Towards generating cytotoxic agents at cancer sites, Br.J.Cancer, 58:700.

Baldwin, R.W., 1985, Monoclonal antibody targeting of anti-cancer agents: Muhlbock Memorial Lecture, Eur.J.Clin.Oncol, 21:1281.

Bennet, H.S., Luft, J.H. and Hampton, J.C., 1959, Morphological classification of vertebrate blood capillaries, Amer.J.Physiol., 196:381.

Buchegger, F., Mach, J-P., Leonnard, P. and Carrel, S., 1986, Selective tumor localisation of radiolabeled anti-human melanoma monoclonal antibody fragments demonstrated in the nude mouse model, Cancer, 58:655.

Buraggi, G.L., Callegaro, L., Mariani, G., Turrin, A., Cascinelli, N., Attilli, A., Bombardierie, G., Terno, G., Plassio, G., Dovis, M., Mazzuca, N., Natal, P.G., Scassellati, G.A., Rosa, U. and Ferrone, S., 1985, Imaging with I-131-labelled monoclonal antibodies to a high molecular weight melanoma-associated antigen in patients with melanoma: efficiency of whole immunoglobulin and its $F(ab')_2$ fragments, Cancer Res., 45:3378.

Carrasquillo, J.A., Krohn, K.A., Beaumier, P., McGuffin, R.W., Brown, J.P., Hellstrom, K.E., Hellstrom, I. and Larson, S.M., 1984, Diagnosis of and therapy for solid tumors with radio-labeled antibodies and immune fragments, Cancer Treat.Rep., 68:317.

Clauss, M.A. and Jain, R.K., 1988, Microvascular permeability and interstitial diffusion coefficients of IgG in normal and neoplastic tissues. Presented at the Second Conference on the Radioimmunodetection and Radioimmunotherapy of Cancer, Princeton, N.J., Sept. 8-10, 1988.

Cobb, L.M. and Humm, J.L., 1986, Radioimmunotherapy of malignancy using antibody targeted radionuclides, Br.J.Cancer, 54:863.

Cobb, L.M., Humphreys, J.A. and Harrison, A., 1987, The diffusion of a tumor-specific monoclonal antibody in lymphoma infiltrated spleen, Br.J.Cancer, 55:53.

Cobb, L.M., Butler, S. and Humphreys, J., 1989, Distribution of injected monoclonal antibody in lymphoma-infiltrated spleen, Leukaemia Res., 13:763.

Covell, D.G., Barbet, J., Holton, O.D., Black, C.D.V., Parker, R.J. and Weinstein, J.N., 1986, Pharmacokinetics of monoclonal immunoglobulin G, $F(ab')_2$ and Fab' in mice, Cancer Res., 46:3969.

Dvorak, H.F., Nagy, J.A., Dvorak, J.T. and Dvorak, A.M., 1988, Identification and characterization of the blood vessels of solid tumors that are leaky to circulating macromolecules, Amer.J.Pathol., 133:95.

Edwards, P.A.W., 1985, Heterogenous expression of cell-surface antigens in normal epithelia and their tumors, revealed by monoclonal antibodies, Br.J.Cancer, 51:149.

Eppenetos, A.A., Munro, A.J., Stewart, S., Rampling, R., Lambert, H.E., McKenzie, C.G., Soutter, P., Rahemtulla, A., Hooker, G., Sivolapenko, G.B., Snook, D., Courtenay-Luck, N., Dhokia, B., Krausz, T., Taylor-Papadimitriou, J., Durbin, H. and Bodmer, W.F., 1987, Antibody-guided irradiation of advanced ovarian cancer with intraperitoneally administered radiolabeled monoclonal antibodies, J.Clin.Oncol., 5:1890.

Gray, H., 1980, Angiology, in: "Gray's Anatomy", P.L. Williams, R. Warwick, eds., 36th ed., Churchill Livingstone, London.

Harwood, P.J., Boden, J., Pedley, R.B., Rawlins, G., Rogers, G.T. and Bagshawe, K.D., 1985, Comparative tumour localisation of antibody fragments and intact IgG in nude mice bearing a CEA-producing human colon tumour xenograft, Eur.J.Cancer Clin.Oncol., 21:1515.

Hirano, A. and Zimmerman, H.M., 1972, Fenestrated blood vessels in a metastatic renal carcinoma in the brain, Lab.Invest., 26:465.

Hori, K., Suzuki, M., Abe, I. and Saito, S., 1986, Increased tumor pressure in association with the growth of rat tumors, Jpn.J.Cancer Res., 77:65.

Humm, J.L. and Cobb, L.M., 1990, Microdosimetry in radioimmunotherapy, J.Nucl.Med., 31, in press.

Iozzo, R.V., 1987, Proteoglycans and the intercellular tumor matrix, Current Topics in Pathology, 77:207.

Jain, R.K., 1987a, Transport of molecules in the tumor interstitium: a review, Cancer Res., 47:3039.

Jain, R.K., 1987b, Transport of molecules across tumor vasculature, Cancer and Metastasis Reviews, 6:559.

Jain, R.K., 1989, Delivery of novel therapeutic agents in tumors: physiological barriers and strategies, J.Natl.Cancer Inst., 81:570.

Karnovsky, M.J., 1968, The ultrastructural basis of transcapillary exchanges, J.Gen.Physiol., 52:641.

Keele, C.A., Neil, E. and Joels, N., eds., 1982, The capillary circulation, in: "Samson Wright's Applied Physiology", 13th ed., Oxford University Press, Oxford.

Ludatscher, R.M., Gellei, B. and Barzilai, A., 1979, Ultrastructural observation on the capillaries of human thyroid tumors, J.Pathol., 138:57.

Mach, J-P., Forni, M., Ritschard, J., Buchegger, F., Carrel, S., Widgren, S., Donath, A. and Alberto, P., 1980, Use and limitations of radiolabeled anti-CEA antibodies and their fragments for photoscanning detection of human colorectal carcinomas, Oncodev.Biol.Med., 1:49.

Mach, J-P., Chatal, J-F., Lumbroso, J-D., Buchegger, F., Forni, M., Ritschard, J., Berche, C., Douillard, J.Y., Carrel, S., Herlyn, M., Steplewski, Z. and Koprowski, H., 1983, Tumor localisation in patients by radiolabeled monoclonal antibodies against colon carcinoma, Cancer Res., 43:5593.

Nugent, L.J. and Jain, R.K., 1984a, Pore and fiber-matrix models for diffusive transport in normal and neoplastic tissues, Microvasc.Res., 28:270.

Nugent, L.J. and Jain, R.K., 1984b, Extravascular diffusion in normal and neoplastic tissues, Cancer Res., 44:238.

Paskins-Hurlburt, A.J., Hollenberg, N.K. and Abrams, H.L., 1982, Tumor perfusion in relation to the rapid growth phase and necrosis: studies on the Walker carcinoma in the rat testicle, Microvasc.Res., 24:15.

Schelin, U., 1962, Chromophobe and acidophil adenoma of the human pituitary gland. A light and electron microscopic study, Acta Pathol. Microbiol.Scand.(Suppl.) 158:5.,

Sunderland, R., Buchegger, F., Schreyer, M., Vacca, A. and Mach, J-P., 1987, Penetration and binding of radiolabeled anti-carcinoembryonic antigen monoclonal antibodies and their antigen binding fragments in human colon multicellular tumor spheroids, Cancer Res., 47:1627.

Swabb, E.A., Wei, J. and Gullino, P.M., 1974, Diffusion and convection in normal and neoplastic tissues, Cancer Res., 34:2814.

Sweet, M.B.E., Thonar, E.J-M.A., Berson, S.D., Skikne, M.I., Immelman, A.R. and Kerr, W.A., 1979, Biochemical studies of the matrix of craniovertebral chordoma and a metastasis, Cancer, 44:652.

Taylor, A.E. and Granger, D.N., 1984, Exchange of macromolecules across the microcirculation, in: "Handbook of Physiology, Section 2: The Cardiovascular System, Vol. IV, Microcirculation", E.M. Renkin and C.C. Michel, eds., American Physiological Society, Bethesda.

Thorpe, P.E., 1985, in: "Monoclonal Antibodies 84: Biological and Clinical Applications", A. Pinchera, G. Doris, F. Dammacco and A. Bargelles, eds., Kurtis, Milan.

Wahl, R.L., Parker, C.W. and Philpott, G.W., 1983, Improved radioimaging and tumor localisation with monoclonal F(ab')$_2$, J.Nucl.Med., 24:316.

Wang, W. and Campiche, M., 1982, Microvasculature of human colorectal epithelial tumors. An electron microscopic study, Virchows Arch.(A), 397:131.

Wiig, H., Tveit, E., Hultburn, R., Reed, R.K. and Weiss, L., 1982, Interstitial fluid pressure in DMBA-induced rat mammary tumors, Scand.J. Clin.Lab.Invest., 42:159.

ANTITUMOR EFFECTS OF SIX RICIN A-CHAIN IMMUNOTOXINS OF POTENTIAL USE IN THE

TREATMENT OF HODGKIN'S DISEASE

Andreas Engert and Philip Thorpe

Drug Targeting Laboratory, Imperial Cancer Research Fund
Lincoln's Inn Fields, London WC2A 3PX, UK

INTRODUCTION

The fate of patients with Hodgkin's disease who relapse or fail to achieve complete remission with first line treatment is ominous. Second line combination chemotherapy can produce good remission rates although cures are uncommon (Hagemeister et al, 1987; Santoro et al, 1986). Of those patients achieving complete remissions, 15-20% will develop a second malignancy as a chemotherapy-related side effect (Valagussa et al, 1986). Therefore new non-mutagenic agents are needed for the treatment of this disease.

One approach would be to link the ribosome damaging A-chain of ricin to antibodies directed against Hodgkin and Reed-Sternberg cell-associated antigens. Since these cells express consistently high levels of the 105/120 KDa CD30 antigen, this antigen is an obvious target for immunotoxins.

Here we describe the antitumor effects of immunotoxins made by linking ricin A-chain to the monoclonal antibodies HRS-1, HRS-3, HRS-4, Ki-1, Ber-H2 which recognize CD30 and to IRac which binds to a different 70 kDa antigen.

MATERIALS AND METHODS

Blue Sepharose CL-6B, Sepharose G25 (fine grade), and Sephacryl S200 HR were obtained from Pharmacia Ltd. (Milton Keynes, England).

Carrier free [^{125}I]Iodine and L-[4,5-^{3}H] leucine (TRK 170) were purchased from Amersham International (Amersham, England). IODO-GEN was from Pierce Ltd. (Chester, England). Silicone fluids Dow Corning 200/1CS, 200/5CS and 550 were purchased from Dow Corning Corp. (Midland, USA).

Cells

The cell line L540, which derived from a patient with Hodgkin's disease (Schaad et al, 1980) was maintained in RPMI 1640 supplemented with 20% (v/v) fetal calf serum, 4 mM L-glutamine, 200 U/ml penicillin and 100 µg/ml streptomycin ('complete medium').

Targeting of Drugs, Edited by G. Gregoriadis *et al.*
Plenum Press, New York, 1990

Antibodies

Five mouse monoclonal antibodies which recognize the CD30 antigen were used in this study: HRS-1 (IgG_{2a}), HRS-3 (IgG_1), HRS-4, (IgG_1), Ki-1 (IgG_3) and Ber-H2 (IgG_1) (Pfreundschuh et al, 1988; Schwab et al, 1982; Schwarting et al, 1987). IRac (IgG_1) recognizes a 70 kDa antigen which is present on Hodgkin and Reed-Sternberg cells as well as on interdigitating reticulum cells (Hsu et al, 1987). The mouse IgG1 monoclonal antibody MRC OX7 which recognizes the mouse Thy 1.1 antigen was used as a non-specific control antibody.

Scatchard Analysis

Monoclonal antibodies were labeled with ^{125}I using the IODO-GEN reagent (Fraker and Speak, 1978). Labeled antibodies at concentrations ranging from 0.25-32 µg/ml were incubated with L540 cells at 4°C. After one h, the cells were separated from the supernatant by centrifugation through silicone fluid (8.8% 0.200/1CS, 7.2% 200/5CS and 84% 550). The radioactivity of the pellet and the supernatant were counted separately. The total amount of radio-labeled specific antibody associated with the cells was corrected by sub-tracting the amount of irrelevant control antibody (OX7) which bound to the cells under the same conditions. The dissociation constant K_d and the number of antibody molecules bound per cell were calculated by the method of Scatchard (1949).

Preparation of Immunotoxins

Chemically deglycosylated ricin-A chain was used for the preparation of immunotoxins. The methods for deglycosylation and purification of A-chain have been described elsewhere (Fulton et al, 1986; Thorpe et al, 1985).

IgG immunotoxins were prepared as described previously by Thorpe et al (1987). In brief, the antibodies were treated with SMPT to introduce an average of 1.7 activated disulphide groups per molecule of antibody. The derivatised protein was separated from unreacted material by gel chromato-graphy on a Sephadex G25 column and mixed with freshly-reduced ricin A-chain. After 72 h, residual activated disulphide groups were inactivated with 0.2 mM cysteine. The immunotoxin preparation was then purified on a Sephacryl S200 HR column equilibrated in 0.1 M sodium phosphate buffer, pH 7.5. Free antibody was removed from the immunotoxin preparation by chroma-tography on a Blue Sepharose CL-6B column, as described by Knowles and Thorpe (1987).

Cytotoxicity Assays

4×10^4 L540 cells in 200 µl complete medium were distributed in the wells of 96-well microtiter plates and were incubated with immunotoxins for 24 h at 37°C in an atmosphere of 5% CO_2 in humidified air. The cells were then pulsed with 1 µCi [^3H]-leucine for 24 h and were harvested onto glass fibre discs. The radioactivity on the discs was measured using a liquid scintillation counter. The [^3H]-leucine incorporation of immunotoxin-treated cultures was compared to that of untreated control cultures to assess cell killing.

Antitumor Experiments in Mice

For the establishment of solid tumors, 2.5×10^7 L540 cells were in-jected subcutaneously into the right posterior gluteal region of 4-6 week old female triple beige (nu/nu-bg/bg-xid) nude mice. When the tumors reached 60-80 mm^3 (approximately 0.5 cm diameter) immunotoxins or antibodies were injected intravenously into the tail vein of the mice. Eight animals

Table 1. Scatchard Analyses of CD 30 Antibodies

Antibody	K_d (nM$\pm$sd)[1]
HRS-1	160$\pm$40
HRS-3	15$\pm$ 3
HRS-4	7$\pm$ 4
Ber-H2	14$\pm$ 4
Ki-1	380$\pm$90

[1] Values of K_d are the arithmetic mean and standard deviation of the results from 3 separate experiments.

were used in each treatment group. The doses injected represented 40% of the LD$_{50}$ (48 μg). Control groups received unconjugated antibodies or PBS. Tumor diameters were recorded twice a week and the tumor volume was calculated as follows:

$$volume = d^2.D.\frac{\pi}{6}$$

d = shorter tumor diameter
D = longer tumor diameter

The antitumor effects of the different treatments were compared by the growth index which is calculated by dividing the mean tumor volume per group 30 days after treatment by the mean tumor volume per group on the day of treatment (day 1).

RESULTS

Scatchard Analyses of the Binding of CD30 Antibodies to L540 Cells

Table 1 summarizes the results of the Scatchard analyses of the binding of the five intact CD30 antibodies tested. HRS-4 had the highest avidity for L540 cells (K_d = 7 nM). Ber-H2 and HRS-3 had the next highest avidity (K_d values of 14 nM and 15 nM respectively). HRS-1 and Ki-1 showed weaker binding with K_d values of 160 nM and 380 nM respectively. The number of antibody molecules bound per cell at saturation was $1.6-1.7 \times 10^6$.

Cytotoxicity In Vitro

As shown in Table 2, IRac.dgA was the most potent immunotoxin. It inhibited the protein synthesis of L540 cells by 50% at a concentration (IC$_{50}$) of 1×10^{-11}M which is similar to the concentration needed for an equivalent effect with ricin itself. The immunotoxins derived from the high affinity CD30 monoclonal antibodies, HRS-3, HRS-4 and Ber-H2, were 9-20 times less potent than IRac.dgA whereas the immunotoxins derived from the low affinity antibodies HRS-1 and Ki-1 were 1000 fold less effective. These cytotoxic effects were specific since the native antibody and OX7.dgA, an immunotoxin that does not bind to L540 cells, were not toxic at 10^{-6} M.

Table 2. Cytotoxicity of Immunotoxins as Tested on L 540 Cells

Material	$IC_{50}(M)$
HRS-1.dgA	$8.0 \pm 2.0 \times 10^{-9}$
HRS-3.dgA	$9.0 \pm 0.8 \times 10^{-11}$
HRS-4.dgA	$1.0 \pm 0.4 \times 10^{-10}$
Ber-H2.dgA	$2.0 \pm 0.5 \times 10^{-10}$
Ki-1.dgA	$1.0 \pm 0.5 \times 10^{-8}$
IRac.dgA	$1.0 \pm 0.2 \times 10^{-11}$
Ricin	$6.0 \pm 2.0 \times 10^{-12}$
OX7.dgA	$>1 \times 10^{-6}$

Values are the arithmetic mean and standard deviation
of at least 3 separate experiments.

Staining of Normal Human Tissue

Immunoperoxidase staining of 28 different human tissues with HRS-3,
HRS-4, Ber-H2 and IRac using standard methods revealed no major crossreactivity of HRS-3, Ber-H2 and IRac. In contrast, HRS-4 stained normal
pancreatic cells strongly (Table 3). This unexpected crossreactivity
probably precludes the use of this antibody as an immunotoxin in patients.

Antitumor Effects on Solid L540 Tumors In Vivo

In Table 4 are listed the detailed results of a series of antitumor
experiments in which HRS-3, Ber-H2 and IRac immunotoxins were administered
at various doses to nude mice bearing solid L540 tumors. Figure 1a shows a
typical experiment in which immunotoxins were administered to mice with
tumors of 60-80 mm^3 (0.5 cm diameter). All three immunotoxins had impressive antitumor activity. IRac.dgA was the most powerful (growth index 0.8),
followed by HRS-3.dgA (growth index 1.4) and lastly Ber-H2.dgA (growth index
4.6). In contrast, the tumors grew progressively in untreated animals
(growth index 9.7) and in animals treated with an immunotoxin of irrelevant
specificity, OX7.dgA (growth index 8.7). The difference in antitumor
activity between IRac.dgA and HRS-3.dgA did not reach statistical significance ($P>0.1$) whereas the difference between IRac.dgA and Ber-H2.dgA was
statistically significant ($P<0.02$). There were several complete remissions:
17/24 for IRac.dgA, 11/16 for HRS-3.dgA and 3/8 for Ber-H2.dgA. Of these,
five IRac.dgA-treated animals and four HRS-3.dgA-treated animals had
relapses after 5-15 days.

It is possible that part of the antitumor activity of the immunotoxins
was mediated through the antibody component alone, since the native antibodies, when administered at doses equivalent to those in the immunotoxins,
appeared slightly to retard tumor growth (Table 4, Figure 1b). The growth
indices in recipients of native IRac and native HRS-3 were both 7.5 and were
not significantly different ($P>0.1$) from the growth index of 8.7 obtained in
untreated mice.

Table 3. Tissue Staining of Four Monoclonal Antibodies Recognising HD/RS Cells

Tissue	HRS-3	HRS-4	Ber-H2	IRac
Adrenal	−	n.d.	−	−
Brain (cortex)	−	n.d.	−	−
Brainstem	−	n.d.	−	−
Breast	−	−	−	−
Cerebellum	−	n.d.	−	−
Cervix	−*)	n.d.*)	−*)	−
Colon	−	−	−	−
Gall bladder	−	n.d.	−	−
Heart	−	−	−	−
Kidney	−	−	−	−
Liver	−	−	−	−
Lung	−*)	−*)	−*)	−
Lymph node	−	−	−	−
Mucosa (nasal)	−	n.d.	−	−
Oesophagus	−	n.d.	−	−
Ovary	−	n.d.	−	−
Pancreas	−	+++	−	−
Parathyroid	−	n.d.	−	−
Prostate	−	−	−	−
Spleen	−	−	−	−
Stomach (antrum)	−	n.d.	−	−
Stomach (body)	−	−	−	−
Testis	−	n.d.	−	−
Thyroid[1]	−	−	−	−
Thyroid[1]	−*)	n.d.	−*)	−
Thyroid (Hashimoto's)	−	n.d.*)	−	−
Tonsils	−*)	−*)	−*)	−
Uterus	−	n.d.	−	−
Vagina	−	n.d.	−	−

1) Autoimmune Thyroiditis
*) Rare cells within lymphoid tissue stain positively

DISCUSSION

The major findings to emerge from this study were: i) four of the six monoclonal antibodies which bind to Hodgkin and Reed-Sternberg cells formed potently and specifically ricin A-chain immunotoxins that were toxic to the L540 Hodgkin cell line in vitro. ii) a single intravenous injection of the most potent immunotoxins, IRac.dgA and HRS-3.dgA, cured 50 and 44% respectively of mice with solid L540 Hodgkin tumors. iii) Since IRac.dgA and HRS-3.dgA combine potent antitumor effects and little crossreactivity with normal human tissue, they could be used for the treatment of patients with Hodgkin's disease.

IRac.dgA, which binds to a 70 KDa antigen on interdigitating reticulum cells and Hodgkin/Reed-Sternberg cells, formed the most powerful immunotoxin in vitro, inhibiting the protein synthesis of L540 cells by 50% at 1×10^{-11} M. The high affinity anti-CD30 antibodies HRS-3, HRS-4 and Ber-H2 (K_d=15 nM, 7 nM and 14 nM respectively) formed potent immunotoxins (IC_{50}= 9×10^{-11}, 1×10^{-10} and 2×10^{-10} M respectively), whereas the low affinity antibodies HRS-1 and Ki-1 (K_d=160 nM and 380 nM, respectively) formed weak immunotoxins (IC_{50}=0.8 and 1.0×10^{-8} M). Since HRS-1, HRS-3, HRS-4 and

Table 4. Treatment of Solid L540 Tumors with Different Immunotoxins

Treatment	Dose (ug Protein)	Average tumor size per group (mm³)			Total	Number of mice		
		Day 1	Day 30	Growth index[1] (day 30/day 1)		CR	Relapse	PR or NR
PBS	–	69	667	9.7 ± 1.6	24[2]	1	–	23
IRac.dgA	48	79	65	0.8 ± 0.2	24[2]	12	5	7
IRac	40	54	407	7.5 ± 1.5	8	1	–	7
HRS-3.dgA	48	99	134	1.4 ± 0.5	16[3]	7	4	5
HRS-3	40	62	464	7.5 ± 1.8	8	1	1	6
Ber-H2.dgA	48	63	291	4.6 ± 1.8	8	3	–	5
Ber-H2	40	61	529	8.8 ± 1.7	8	1	–	7
OX7.dgA	48	68	589	8.7 ± 1.1	8	–	–	8

1) arithmetic mean ± one standard error
2) average of three separate experiments
3) average of two separate experiments

CR = complete remission
PR = partial remission
NR = no response

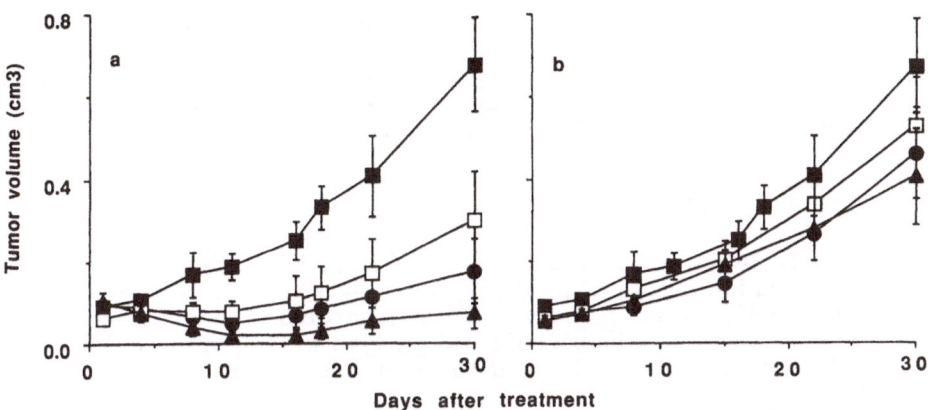

Fig. 1. Antitumour effects of immunotoxins (a) and antibodies (b) on
solid L540 tumors in nude mice. Tumors were approximately 0.5 cm
on the day of treatment (day 1). Groups of 8-10 animals received
i.v. injections of the following: a) ■ PBS, □ Ber-H2.dgA, ● HRS-
3.dgA, ▲ IRac.dgA. b) ■ PBS, □ Ber-H2, ● HRS-3, ▲ IRac. The doses
in terms of protein were 48 µg for immunotoxins and 40 µg for antibodies. Bars represent the standard error of the mean tumor diameter at various times after treatment.

Ber-H2 bind to the same epitope on the CD30 antigen (Engert et al, 1990) it
can be deduced that the affinity of these MoAbs determines their potency as
ricin A-chain immunotoxins.

Despite the numerous reports of immunotoxins with high cytotoxic
potency and specificity in vitro, there have been relatively few reports of
their successful use to treat solid tumors in vivo: Bernhard et al (1983)
and Hwang et al (1984) reported the complete abolition of L10 hepatocarcinoma tumors in guinea pigs by single i.v. or s.c. injections of abrin A-
chain immunotoxins. Leonhard et al (1988) reported 11/46 complete remissions of solid human T cell tumors in mice intravenously treated with CD5
ricin A-chain immunotoxins.

The immunotoxins used in the present study gave impressive antitumor
effects in a solid Hodgkin's disease xenograft model. The growth index
(ratio of average tumor volume on day 30:day 1) was 0.8 for IRac.dgA, 1.4
for HRS-3.dgA and 4.6 for Ber-H2.dgA as compared with 9.7 for PBS treated
control animals. Possible explanations for the high in vivo efficacy of the
present immunotoxins are the use of deglycosylated ricin A-chain (Blakey et
al, 1987), the SMPT linker (Thorpe et al, 1987), and a final purification
step on Blue Sepharose (Knowles and Thorpe, 1987). These procedures, when
combined, give "second generation" immunotoxins that have higher purity,
better in vivo stability, and which avoid liver entrapment better than their
predecessors, resulting in substantially improved antitumor activity in
mouse tumor models (Thorpe et al, 1988).

The CD30 antigen seems to be an ideal target for immunotherapy since it
is strongly expressed in all cases of Hodgkin's disease (Chittal et al,
1988) and has very limited expression in normal tissue (Stein et al, 1985).
This is reflected by the finding that only one of the five CD30 MoAbs tested
(HRS-4) crossreacts with a vital organ (pancreas) that would preclude its
clinical use as an immunotoxin.

In summary, we have demonstrated that immunotoxins made by linking de-glycosylated ricin A-chain to monoclonal antibodies recognizing Hodgkin and Reed-Sternberg cell-associated antigens are powerful and specific antitumor reagents _in vitro_ and _in vivo_. Thus, HRS-3.dgA and IRac.dgA are candidates for the treatment of Hodgkin's disease in man.

REFERENCES

Bernhard, M.I., Foon, K.A., Oeltmann, T.N., Key, M.E., Hwang, K.M., Clarke, G.C., Christensen, W.L., Hoyer, L.C., Hanna, M.G., Jr. and Oldham, R.K., 1983, Guinea pig line 10 hepatocarcinoma model: characterization of monoclonal antibody and _in vivo_ effect of unconjugated antibody and antibody conjugated to diphtheria toxin A-chain, Cancer Res. 43:4420.

Blakey, D.C., Watson, G.J., Knowles, P.P. and Thorpe, P.E., 1987, Effect of chemical deglycosylation of ricin A-chain on the _in vivo_ fate and cytotoxic activity of an immunotoxin composed of ricin A-chain and anti-thy 1.1. antibody, Cancer Res., 47:947.

Chittal, S.M., Caveriviere, P., Schwarting, R., Gerdes, J., Al Saati, T., Rigal-Huguet, F., Stein, H. and Delsol, G., 1988, Monoclonal antibodies in the diagnosis of Hodgkin's disease. The search for a rational panel, Am.J.Surg.Path., 12:9.

Engert, A., Burrows, F., Jung, W., Tazzari, P.L., Stein, H., Pfreundschuh, M., Diehl, V. and Thorpe, P., 1990, Evaluation of ricin A-chain-containing immunotoxins directed against CD30 as potential reagents for the treatment of Hodgkin's disease, Cancer Res., in press.

Fraker, P.J. and Speak, J.C., 1978, Protein and cell membrane iodination with a sparingly soluble chloramide, 1,3,4,6-tetrachloro-3 ,6 diphenylglycouril, Biochem.Biophys.Res.Comm., 80:849.

Fulton, R.J., Blakey, D.C., Knowles, P.P., Uhr, J.W., Thorpe, P.E. and Vitetta, E.S., 1986, Purification of ricin Al, A2 and B chains and characterization of their cytotoxicity, J.Biol.Chem., 261:5314.

Hagemeister, F.B., Tannir, N., McLaughlin, P., Salvador, P., Riggs, S., Velasquez, W.S. and Cabanillas, F., 1987, MIME chemotherapy (Methyl-GAG, Ifosfamide, Methothrexate, Etoposide) as treatment for recurrent Hodgkin's disease, Clin.Oncol., 5:556.

Hsu, S-M., Ho, Y-S. and Hsu, P-L., 1987, Effect of monoclonal antibodies anti-2H9, anti-IRac, and anti-HeFi-1 on the surface antigens of Reed-Sternberg cells, J.Natl.Cancer Inst., 5:1091.

Hwang, K.M., Foon, K.A., Cheung, P.H., Pearson, J.W. and Oldham, R.K., 1984, Selective antitumor effect on L10 Hepatocarcinoma cells of a potent immunoconjugate composed of the A-chain of abrin and a monoclonal antibody to a hepatoma-associated antigen, Cancer Res., 44:4578.

Knowles, P.P. and Thorpe, P.E., 1987, Purification of immunotoxins containing ricin A-chain and abrin A-chain using Blue Sepharose CL-6B, Anal.Biochem., 160:440.

Leonard, J.E., Johnson, D.E., Shawler, D.L. and Dillman, R.O., 1988, Inhibition of human T-cell tumor growth by T101-ricin A-chain in an athymic mouse model, Cancer Res., 48:4862.

Pfreundschuh, M., Mommertz, E., Meissner, M., Feller, A.C., Hassa, R., Krueger, G.R.F. and Diehl, V., 1988, Hodgkin and Reed-Sternberg cell associated monoclonal antibodies HRS-1 and HSR-2 react with activated cells of lymphoid and monocytoid origin, Anticancer Res., 8:217.

Santoro, A., Viviani, P., Valagussa, P., Bonfante, V. and Bonadonna, G., 1986, CCNU, Etoposide and Prednimustine (CEP) in refractory Hodgkin's disease, Semin.Oncol., 13(Suppl. 1):23.

Scatchard, G., 1949, The attraction of proteins for small molecules and ions, Ann.N.Y.Acad.Sci., 51:660.

Schaadt, M., Diehl, V., Stein, H., Fonatsch, C. and Kirchner, H.H., 1980,

Two neoplastic cell lines with unique features derived from Hodgkin's disease, Int.J.Cancer, 26:723.

Schwab, U., Stein, H., Gerdes, J., Lembke, H., Kirchner, H., Schaadt, M. and Diehl, V., 1982, Production of a monoclonal antibody specific for Hodgkin and Sternberg-Reed cells of Hodgkin's disease and a subset of normal lymphoid cells, Nature, 299:65.

Schwarting, R., Gerdes, J. and Stein, H., 1987, Ber-H2: a new monoclonal antibody of the Ki-1 family for the detection of Hodgkin's disease in formaldehyde-fixed tissue sections, in: "Leucocyte Typing III", J.A. McMichael, ed., Oxford University Press, Oxford.

Stein, H., Mason, D.Y., Gerdes, J., O'Connor, N., Wainscoat, J., Pallesen, G., Gatter, K., Falini, B., Delsol, G., Lembke, H., Schwarting, R. and Lennert, K., 1985, The expression of the Hodgkin's disease associated antigen Ki-1 in reactive and neoplastic lymphoid tissue: evidence that Sternberg-Reed cells and histiocytic malignancies are derived from activated lymphoid cells, Blood, 66:848.

Thorpe, P.E., Detre, S.I., Foxwell, B.M.J., Brown, A.N.F., Skilleter, D.N, Wilson, G., Forrester, J.A. and Stirpe, F., 1985, Modification of the carbohydrate in ricin with metaperiodate-cyanoborohydride mixtures. Effect on toxicity and in vivo distribution, Eur.J.Biochem., 147:197.

Thorpe, P.E., Wallace, P.M., Knowles, P.P. Relf, M.G., Brown, A.N.F., Watson, G.J., Knyba, R.E., Wawrzynczak, E.J. and Blakey, D.C., 1987, New coupling agents for the synthesis of immunotoxins containing a hindered disulfide bond with improved stability in vivo, Cancer Res., 47:5924.

Thorpe, P.E., Wallace, P.M., Knowles, P.P., Relf, M.G., Brown, A.N.F., Watson, G.J., Blakey, D.C. and Newell, D.R., 1988, Improved anti-tumor effects of immunotoxins prepared with deglycosylated ricin A-chain and hindered disulfide linkages, Cancer Res., 48:6396.

Valagussa, P., Santora, A., Fossati-Belloni, F., Banti, A. and Bonadonna, G., 1986, Second acute leukemia and other malignancies following treatment for Hodgkin's disease, J.Clin.Oncol., 4:830.

ENHANCEMENT OF HORMONE ACTIVITY BY MONOCLONAL ANTIBODIES

R. Bomford and R. Aston*

Wellcome Biotechnology, Langley Court, Beckenham, Kent
BR3 3BS, UK. *Peptide Technology Ltd., P.O. Box 444
Dee Why, NSW 2099, Australia

INTRODUCTION

The exploitation of genetic engineering technology has made available
many protein hormones and mediators for clinical application in man; however, it is becoming progressively apparent that some of these recombinant
molecules (e.g. TNF, IL-1, IL-2) retain high levels of toxicity in vivo. As
a result of such observations, there is increasing interest in ways of
improving the biological half-life and specificity of potential therapeutic
agents. At the same time, it is also becoming apparent that many hormones/
mediators recognise several receptor subtypes distinguished by properties
such as structure, affinity, specificity and tissue origin. It is probable
that the different cellular receptors for hormones have evolved in order to
provide an additional step in the complex regulatory process leading to a
physiological response. Recent evidence also suggests some cellular receptors are released into the circulation in order to act as "carrier" proteins
for hormones. Whether this effect is important to protect the relevant
hormone from degradation or to provide a more optimal delivery and function
is as yet unclear.

This paper describes the enhancement of the biological activity of
growth hormone (GH) in vivo by particular site directed antisera or monoclonal antibodies (MAbs). This phenomenon appears to involve the "targeting" of hormones to receptors of particular sub-type, since it has been
shown that certain hormone-MAb complexes will only interact with certain
receptors. The enhancement of the biological activity of GH by MAbs was
first discovered with human GH (Holder et al, 1985; Aston et al, 1986) and
subsequently confirmed with bovine GH (Aston et al, 1987) and the general
area of antibody-mediated enhancement has recently been reviewed (Aston et
al, 1989).

GHs are produced in the pituitary and are proteins of about 190 amino
acids, the length varying slightly between species (Miller and Eberhardt,
1983). The structure of pig GH has been solved by X-ray crystallography and
has been found to be an alpha-helical bundle, comprising four alpha-helices
joined together by loops (Abdel-Meguid et al, 1987). The growth-promoting
activity of GH is not mediated directly, but via the stimulation of the
release of insulin-like growth factors (otherwise called somatomedins) of
which the liver is a particularly rich source (Cadman and Wallace, 1981).

These in turn induce mitogenicity and growth of various tissues, including cartilage. The assay system that was employed for the work described below was the uptake of $^{35}SO_4^{2-}$ into the costal cartilage of dwarf mice, which are characterised by hypoplasia of the anterior region of the pituitary gland and are deficient in GH (Holder et al, 1980). The correlation of this metabolic parameter with somatic growth has been well established (Aston et al, 1987).

ENHANCEMENT OF GH ACTIVITY

MAbs against bovine GH enhance the uptake of $^{35}SO_4^{2-}$ into the costal cartilage of dwarf mice and the effect depends on both the dose of MAb and hormone. Figure 1 shows a titration of bovine GH in the presence or absence of a fixed dose of MAb. It may be noted that the growth rate observed with 80 microgrammes of bovine GH on its own was less than that achieved with 5 microgrammes in complex with antibody, representing an enhancement of hormone potency of over fifteen-fold. When the effect of seven different MAbs was examined (Fig. 2) it was found that five of them significantly potentiated activity, whereas the remaining two were without effect. The correlation between growth rate and IGF_1 levels during the course of antibody-mediated enhancement has been demonstrated elsewhere (Wallis et al, 1987). No MAbs which blocked activity in this in vivo system were detected, although antibodies against human GH which potentiated hormone activity in vivo also inhibited its growth promoting effect on a lymphome cell line in vitro (Aston et al, 1986). This underlines the important point that the

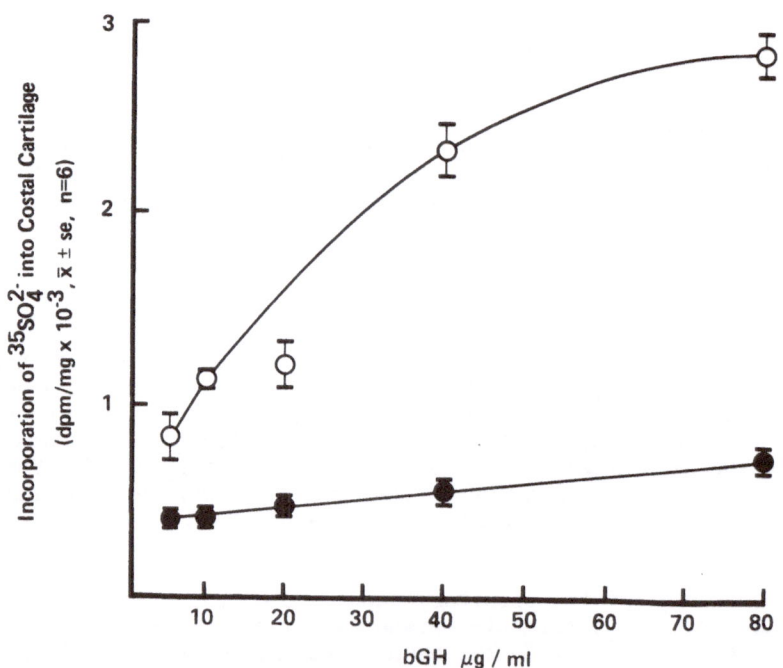

Fig. 1.   Enhancement of bovine GH activity in vivo by MAb OA11. Growth rate of dwarf mice was determined by $^{35}SO_4^{2-}$ incorporation into costal cartilage in the presence of varying doses of GH either free (●——●) or in complex with OA11 (○——○) (reproduced from Aston et al, 1987, with permission).

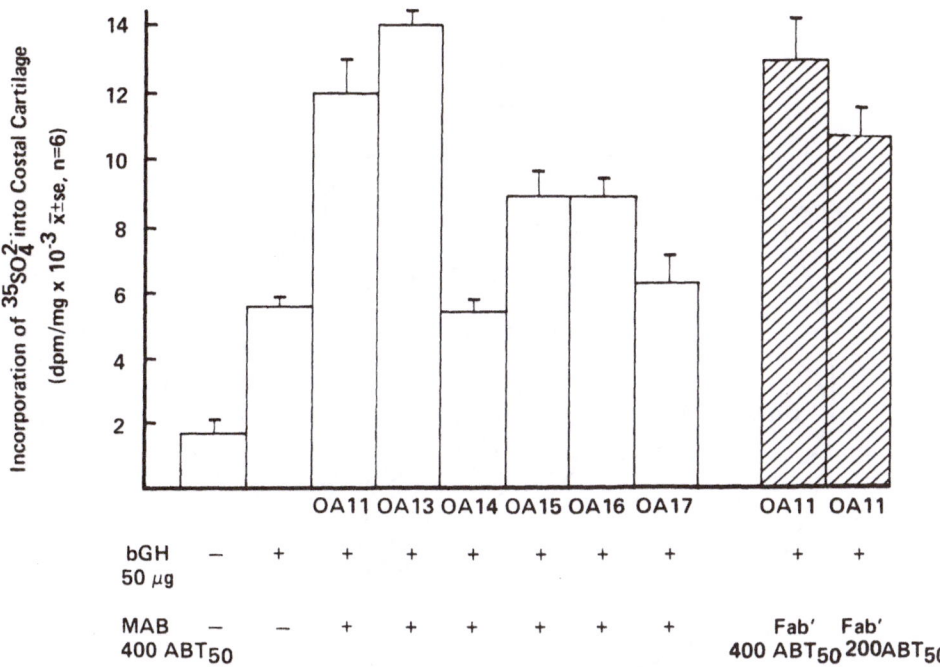

Fig. 2. Effects of MAbs of different epitope specificity on bovine GH
activity _in vivo_. Dwarf mice were injected with GH either free or
in complex with different MAbs (OA11-17). Growth rate was determined by $^{35}SO_4^{2-}$ incorporation into costal cartilage. Hatched bars
correspond to enhancement of GH activity by univalent Fab' fragments of MAb OA11 (reproduced from Aston et al, 1987, with permission).

effect of an anti-GH antibody on hormone activity may depend on the assay
system employed, which needs to be borne in mind when considering alternative hypotheses for the enhancement phenomenon. Clearly, certain mechanisms can only be operative in an _in vivo_ system, for example where inactivation may occur through serum enzymes or particular hormones have
targets on a variety of tissues exhibiting different receptors. Figure 2
also shows that the enhancement of GH activity can be mediated by monovalent
Fab' fragments of a MAb, a result which indicates that neither bivalency nor
Fc-region-mediated targeting are involved in the enhancement phenomenon.
The number of separate epitopic sites recognised by the MAbs against bovine
GH was determined by competition analysis, the results of which are schematically summarised in Figure 3. The antibodies which enhance hormone
activity (OA11, OA12, OA13, OA15 and OA16) react with three distinguishable
sites, indicating that enhancement does not depend on antibody binding to a
unique site. The location of the antibody binding sites on the bovine GH
molecule has not yet been determined. The four MAbs OA11, OA12, OA13 and
OA14 showed mutual partial cross-competition, suggesting that they are
recognizing topographically related but distinct sites on a major antigenic
determinant. Three of these antibodies (OA11, OA12 and OA13) were enhancing, whereas the fourth (OA14) was not, showing that the ability of an antibody to enhance may be controlled by quite fine changes in specificity.

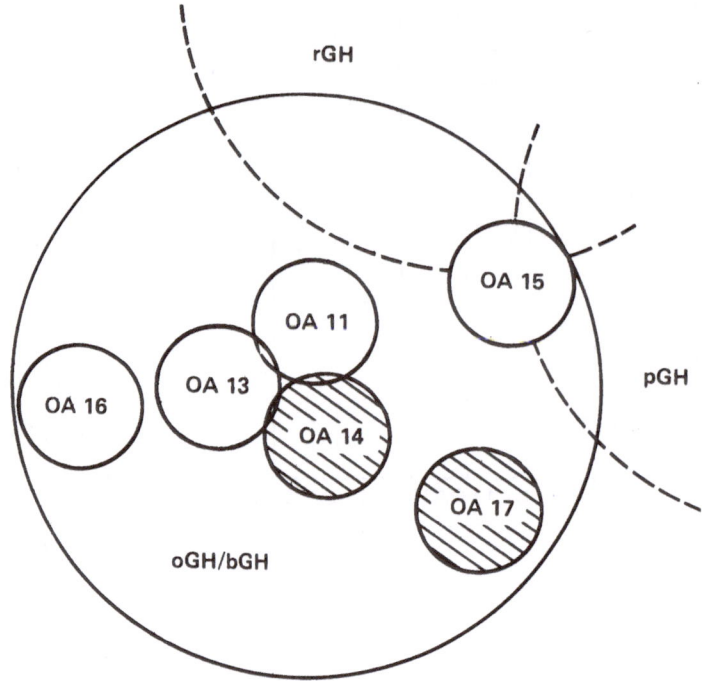

Fig. 3. Schematic representation of antigenic determinants on bovine GH
defined by MAbs. Overlap between circles (epitopes) indicates
effective competition between pairs of antibodies for binding to
GH. Antibody OA15 cross-reacts with pig GH (pGH) and rat GH (rGH)
(reproduced from Aston et al, 1987, with permission).

HYPOTHESES FOR THE MECHANISM OF ENHANCEMENT BY MONOCLONAL ANTIBODIES

Cross-linking of Hormone Receptors

The cross-linking of cell surface receptors is known to be an important
aspect of the induction of hormone action in vitro. It has also been
established that certain antisera to receptors of, for instance, insulin can
'mimic' hormone action by facilitating the cross-linking of cell surface
receptors. This phenomenon is effected through the bivalent nature of antibody molecules and it has been shown that univalent antibody fragments are
without activity. The enhancement of hormonal activity by this mechanism,
for insulin and EGF, has been achieved by employing selected antibodies to
these hormones (Shechter et al, 1979).

Targeting to Fc Receptors in the Liver

The fact that liver is a rich source of both GH and Fc receptors and a
major site for IGF_1 synthesis, raises the possibility that Fc-region mediated targeting of GH-MAb complexes was occurring in this organ. However,
the observation that univalent FAb' fragments derived by papain digestion of
antibody, were equipotent to intact bivalent antibody, excludes both the
above mechanisms.

Increased Plasma Half-life

The complexes of GH and MAb are soluble and therefore could have an
extended half-life in the circulation; however, a role for this mechanism in

MAb mediated GH enhancement is unlikely from observations which indicate that enhancement could be achieved by a non-systemic bioassay for lactogenic activity (Aston et al, 1986). This assay involves local administration of the treatment material adjacent to the crop sac, the two halves of which may be treated independently, showing that systemic effects are not operating. Despite these observations Clark et al (1985) show that continuous or pulsatile administration of GH results in marked enhancement of biological activity in comparison to a bolus injection.

Allosteric Modification of GH

This is a mechanism that is believed to play a role in the various effects that MAbs exert on the enzymic activity of beta-galactosidase (Frackelton and Rotman, 1980). In the case of GH, it would be expected that an allosteric effect induced by an enhancing MAb should increase the affinity of binding of GH to its receptor. In vitro studies of the effect of MAbs on the binding of bovine GH to receptors on microsomes prepared from sheep or rabbit liver appeared to offer some evidence against this, in so far as one of the enhancing antibodies (OA16) inhibited receptor binding (Aston et al, 1987). However, this result is not conclusive in ruling out the hypothesis of allosteric modification because, as explained more fully below, there are probably multiple different receptors for GH, and the in vitro studies may not have been measuring binding to the one which mediates the relevant biological activity.

Receptor Targeting

It is a prerequisite for this hypothesis that there are several subtypes of GH receptor and that the regions on the GH molecule involved in binding to these receptors are topographically distinct. Thus, a MAb which blocks the binding site for one type of receptor need have no effect on (or may even potentiate) binding to alternative receptors. The multiple receptor hypothesis explains why a single MAb can display an enhancing effect in one biological assay and yet inhibit activity in a second. The circumstantial evidence in favour of this hypothesis has been reviewed extensively elsewhere (Aston et al, 1989) and here it only needs to be added that the existence of different GH receptors is now an established fact, the genes for two different receptors having been cloned and sequenced (Smith et al, 1988).

In summary, the mechanism(s) of the enhancement of GH activity by MAbs remains undetermined, although any which depend on antibody divalency or the Fc fragment can be ruled out. It should be stressed that there is no need to assume that only a single mechanism is operating; it is quite feasible that alternative mechanisms could be mediated by the same or different antibodies.

CONCLUSIONS

The study of the effect of MAbs on the biological activity of GH has revealed that they can block or promote its activity, depending on the assay system and the site on the hormone recognised by the antibody. At this stage, it is not clear whether or not all the biological activities of GH are similarly enhanced by antibody. If antibody mediated enhancement of hormonal activity is restricted to particular effects or properties of the hormone, this approach may lend itself to the improvement of the therapeutic index of other hormones or mediators. It will be extremely interesting to find out if similar observations will be made with the increasing numbers of protein mediators (cytokines) that are being discovered to play an important role in the immune and inflammatory systems. These molecules are well known to manifest both desirable and detrimental activities. Indeed, the enhancement of the biological activity of thyroid stimulating hormone (TSH) (Holder et al, 1987) and gamma-interferon (Schreiber et al, 1985) by MAbs has already been described. If the phenomenon does turn out to be general, what potential problems or therapeutic opportunities does it present?

Up to now, the efforts that have been made to intervene therapeutically by immunological means in the biological activity of hormones or cytokines have had the aim of blocking biological activity. Examples include active immunization with the hormone chorionic gonadotrophin to restrict fertility (Jones et al, 1988) or passive immunization with antibodies against tumour necrosis factor (TNF) to control septic shock (Beutler et al, 1985). In such situations where the antibodies are intended to block biological activity, it is necessary to be aware of the possibility of enhancement, which might arise because the specificity of actively-induced antibody could change between individuals. The consequences of a passively administered monoclonal antibody may be variable due to the inhibition of one activity, whilst at the same time a second biological activity may be enhanced.

Another treatment which causes the appearance of anti-hormone antibodies, albeit without deliberate intent, is hormone replacement therapy with insulin or GH. There is a considerable amount of indirect evidence that these antibodies may in many, if not most, cases be inconsequential or enhance rather than block hormone activity (Aston et al, 1989). It seems that in these cases, the antibodies could be conferring an inadvertent clinical benefit through their enhancing properties. It is clearly very difficult to provide direct evidence for the enhancing effects of antibodies in patients; however, a study of five patients with familial GH deficiency and high levels of circulating antibody has shown that one of these retained specificities which on passive transfer to hypopituitary mice significantly enhanced growth rate (Aston et al, in preparation).

The observation that hormonal enhancement is an epitope-related effect raises fundamental questions as to the role of antibodies appearing during the course of hormone replacement therapy. Although polyclonal in nature, such antisera are likely to be of highly restricted epitope specificity due to the fact that the antigen is "self" or highly homologous to the endogenous hormone (e.g. human GH or porcine insulin). Under such conditions it may be expected that the patient would manifest one of three responses to the hormone therapy: (i) refractoriness to treatment as a result of the production of inhibitor antisera; (ii) no effect; or (iii) enhancement. In the case of the latter effect, mechanisms such as facilitated cross-linking of receptors or alteration of the pharmacokinetics of the hormone may influence the overall effect. The polyclonal nature of autoimmune antisera to hormones also dictates that there may be combinations of enhancing and inhibitory specificities present. The overall effect of the immune response in a patient to a therapeutic hormone would be dependent upon a number of properties of the antiserum; these include: (i) specificity; (ii) polyclonality; (iii) whether or not the antiserum causes insolubilization of the antigen; (iv) rate at which complexes are cleared by the reticuloendothelial system; and (v) prolongation of serum half-life or alteration of pharmacokinetics of the hormone.

In what ways could antibody-mediated enhancement be exploited to deliberate advantage? In the specific case of GH this is already being explored as a potential means of promoting the growth of farm livestock. Sheep which are auto-immunised with selected synthetic peptides from ovine GH develop antibodies of restricted specificity which enhance the biological activity of GH in the dwarf mouse model (Bomford and Aston, 1990; Aston et al, 1990) raising the possibility of a growth promotion vaccine. There would also appear to be scope for the application of passive administration of enhancing monoclonals to improve cytokine therapy in humans, although this would require the humanisation of rodent monoclonals.

REFERENCES

Abel-Meguid, S.S., Shieh, H-S., Smith, W.W., Dayringer, H.E., Violand, B.N. and Bentle, L.A., 1987, Three dimensional structure of a genetically engineered variant of porcine growth hormone, Proc.Natl.Acad.Sci., USA, 84:6434.

Aston, R., Holder, A.T., Preece, M.A. and Ivanyi, J., 1986, Potentiation of the somatogenic and lactogenic activity of human growth hormone with monoclonal antibodies, J.Endocrinol., 110:381.

Aston, R., Holder, A.T., Ivanyi, J. and Bomford, R., 1987, Enhancement of bovine growth hormone activity in vivo by monoclonal antibodies, Molec.Immunol., 24:143.

Aston, R., Cowden, W.B. and Ada, G.L., 1989, Antibody-mediated enhancement of hormone activity, Molec.Immunol., 26:435.

Aston, R., Rathjen, D.A., Holder, A.T., Bender, V., Trigg, T.E., Cowan, K., Edwards, J.A. and Cowden, W.B., 1990, Antigenic structure of bovine growth hormone: location of a growth enhancing region, (submitted for publication).

Beutler, B., Milsark, I.W. and Cerami, A.C., 1985, Passive immunization against cachectin/tumor necrosis factor protects mice from lethal effect of endotoxin, Science, 229:869.

Bomford, R. and Aston, R., 1990, Enhancement of bovine growth hormone activity by anti-peptide auto-antibody, J.Endocrinol., in press.

Cadman, H.F. and Wallis, M., 1981, An investigation of the sites that bind human somatotropin (growth hormone) in the liver of the pregnant rabbit, Biochem.J., 198:605.

Clark, R.G., Jansson, J-O., Isakson, O. and Robinson, I.C.A.F., 1985, Intravenous growth hormone: growth responses to patterned infusion in the hypophysectionized rat, J.Endocrinol., 104:53.

Frackelton, A.R. and Rotman, B., 1980, Functional diversity of antibodies elicited by bacterial beta-D-galactosidase, J.Biol.Chem., 225:5286.

Holder, A.T., Wallis, M., Biggs, P. and Preece, M.A., 1980, Effects of growth hormone, prolactin and thyroxine on body weight, somatomedin-like activity and in vivo sulphation of cartilage in hypopituitary dwarf mice, J.Endocrinol., 85:35.

Holder, A.T., Aston, R., Preece, M.A. and Ivanyi, J., 1985, Monoclonal antibody-mediated enhancement of growth hormone activity in vivo, J.Endocrinol., 107:59.

Holder, A.T., Aston, R., Rest, J.R., Hill, D.J., Patel, N. and Ivanyi, J., 1987, Monoclonal antibodies can enhance the biological activity of thyrotrophin, Endocrinology., 120:567.

Jones, W.R., Bradley, J., Judd, S.J., Denholm, E.H., Ing, R.M.Y., Mueller, U.W., Powell, J., Griffin, P.D. and Stevens, V.C., 1988, Phase I clinical trial of a World Health Organisation birth control vaccine, Lancet, i:1295.

Miller, W.L., and Eberhardt, N.L., 1983, Structure and evolution of the growth hormone gene family, Endocrine Rev., 4:97.

Schreiber, R.D., Hicks, L.J., Celada, A., Buchmeier, N.A. and Gray, P.W., 1985, Monoclonal antibodies to murine gamma-interferon which differentially modulate macrophage activation and antiviral activity, J.Immunol., 134:1609.

Shechter, Y., Chang, K.J., Jacobs, S. and Cuatrecasas, P., 1979, Modulation of binding and bioactivity of insulin by anti-insulin antibody: relation to possible role of receptor self aggregation in hormone action, Proc.Natl.Acad.Sci.USA, 76:2720.

Smith, W.C., Linzer, D.I. and Talamantes, F., 1988, Detection of two growth hormone receptor mRNAs and primary translation products in the mouse, Proc.Natl.Acad.Sci.USA, 85:9576.

Wallis, M., Daniels, M., Ray, K.P., Cottingham, J.D. and Aston, R., 1987, Monoclonal antibodies to bovine growth hormone potentiate effects of the hormone on somatomedin C levels and growth of hypophysectionized rats, Biochem.Biophys.Res.Comm., 149:187.

MODELING OF CELL MEMBRANE TARGETING: SPECIFIC RECOGNITION, BINDING, AND PROTEIN DOMAIN FORMATION IN LIGAND-CONTAINING MODEL BIOMEMBRANES

D.W. Grainger*, M. Ahlers[#], R. Blankenburg[#], P. Meller[#]
A. Reichert[#], H. Ringsdorf[#] and C. Salesse**

*Department of Chemical and Biological Sciences, Oregon Graduate Institute of Science and Technology, Beaverton OR 97006-1999, USA. [#]Institut für Organische Chemie Universität Mainz, D-6500 Mainz, West Germany. **Centre de Recherche en Photobiophysique, Université Quebec, Trois Rivieres, Quebec, Canada

INTRODUCTION

Drug delivery systems are designed to assist, accelerate, and control transport of pharmacologically active agents from sites of administration to specified targets in organs and tissues. So-called controlled drug delivery systems are intended to maintain continuously efficacious drug concentrations in vivo, either locally or systemically, over longer time periods. They should provide constant dosage levels above a minimum level of efficacy yet below mandated toxicity levels - a significant advantage over many conventional systemically administered formulations. Site-specific targeting of drugs, particularly those agents which prove highly toxic in small doses, can be utilized to maintain therapeutically relevant levels of drug to targeted tissue at a localized site specifically without systemic toxicity. Pharmaceutical problems, such as avoidance of first-pass effects in the liver, spleen, and filtration organs, solubility and stability problems in certain formulations, as well as the reliability of long-term delivery within a single device are important advantages in clinical applications of controlled and targeted drug delivery systems.

However, as in most scientific problems, surmounting practical obstacles in drug delivery has been much more difficult than formulating strategies. In fact, the general problems surrounding effective controlled and targeted drug delivery are long standing: for a bioactive agent to elicit its desired response, it must locate its site of action more specifically, faster, with higher affinity, and with greater probability than its being degraded, displaced, excreted, inactivated, or nonspecifically adsorbed or absorbed. Between the pharmacon and the site of action, there exists a multitude of barriers to penetrate (gut, RES, cell membrane) before the drug successfully reaches its proper destination - factors which must be accounted for in designs of modern delivery systems.

Most targeting strategies presume the existence of a feature on the cell membrane exterior that is a focal point for aiming the carrier system. To compliment the target, these strategies often utilize a highly selective molecule able to recognize this target feature and to deliver a drug to the designated goal. Often, the target is a cell membrane epitope, receptor, or distinct molecular feature attached or integral to the lipid matrix of a cell membrane. Moreover, antibodies or significant fragments of antibodies are utilized to specifically recognize only these targets and bind with high affinity to the cell membrane at these sites. In fundamental terms, the key step to targeted delivery is often simply a specific recognition event between the delivery vehicle (typically a protein) and a cellular target. The recognition event itself is often implicit in the success of a given strategy; in fact, it is rarely discussed in mechanistic terms and is often simply assumed to occur. However, insight into the physical chemistry of target recognition could benefit future targeting strategies in terms of allowing design of specific delivery vehicles for targeted ligands, and modeling the processes involved for any vehicle to effectively access and bind a target feature.

Cell membrane mimicry (Ringsdorf et al, 1988 and references therein) has improved the understanding, both physically and chemically, of the nature of the barrier structure that the cell membrane presents. Model systems (Ringsdorf et al, 1988; Gregoriadis and Allison, 1980; Mohwald, 1988; Dhathathreyan et al, 1988; Thompson et al, 1988) allow simplification of an otherwise impossibly complex cell membrane: monolayers, that despite their simplicity, can reveal much about lipid packing and lateral organization, as well as solute (drug) and protein association in membrane bilayers (Dhathathreyan and Mobius, 1988; Ibdah and Phillips, 1988; Heckl et al, 1987; Grainger et al, 1989a), both planar and as vesicles, which allow measurements of membrane ordering and permeability (Mayer et al, 1988; Venema and Weringa, 1988; Chen et al, 1988), reconstitution of active membrane channels, transport proteins, and lectins for surface recognition reactions (Ohtoyo et al, 1988; Fendler et al, 1987), assessment of dynamic membrane process (Sackmann et al, 1986; Gaub et al, 1985), and applications

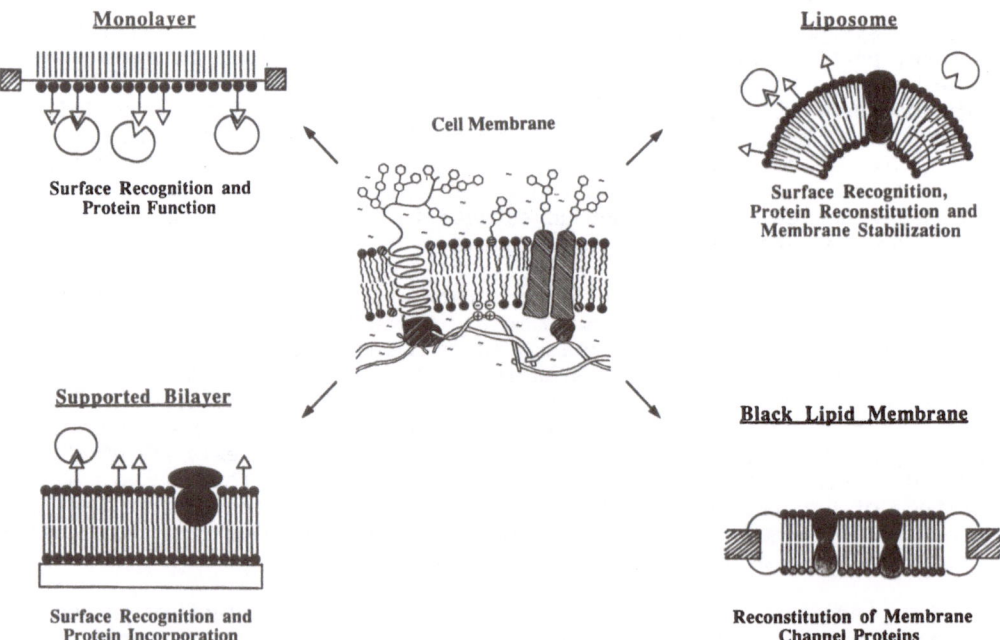

Fig. 1. Various model membrane systems available for cell membrane investigations.

as drug carriers (Gregoriadis and Allison, 1980; Bundgaard et al, 1981). These experimental systems are schematically shown in Fig. 1 as model bio-membranes.

In liposomal drug delivery systems in particular, two possible formulations are (a) encapsulation of drugs within the aqueous interior of lipid vesicles, or (b) liposomes from lipids with pharmacologically active head groups (Hashimoto et al, 1985; Matsuhita et al, 1981; Rosemeyer et al, 1985). The latter form yields a novel prodrug which may be inactive during transport, improves drug solubility by lipid conjugation, and provides direct interaction of drug with cell membrane targets through direct lipid exchange or endocytosis. Such a delivery system strategy demonstrates both capabilities for drug function and target recognition.

The purpose of this communication is to present two different biophysical studies that successfully model surface recognition and binding of proteins (as ligands) to specific receptors or targets bound in model biomembranes. In the first study, the specific binding of subphase-solubilized streptavidin with monolayers of different lipids with biotin (vitamin H) has been discriminated from nonspecific adsorption with surface pressure isotherms. Additionally, through specific recognition and binding to monolayers of biotin lipids, fluorescently labeled streptavidin spontaneously organizes in the plane of the monolayer to create large, regular proteinaceous domains at the interface, directly visible in situ by epifluorescence microscopic observation at the gas-water interface (Blankenburg et al, 1989). These domains have been transferred onto solid supports (Ahlers et al, 1990a) and found to be two-dimensional protein crystals by electron crystallography (Ahlers et al, 1990b). In the second study, the hydrolytic action of the membrane-active enzyme, phospholipase A_2 (PLA$_2$) in monolayers of phospholipids has been characterized. PLA$_2$, a small, ubiquitous enzyme hydrolytically active against phospholipids, has often been utilized previously as a membrane probe and as a model system for enzyme-substrate interactions. Its catalytic hydrolysis of acyl chains from lipid substrates remains of interest for two basic reasons: (a) PLA$_2$'s activity against organized lipid substrates (micelles, monolayers, vesicles) is three to four orders of magnitude greater than that against dispersed monomeric substrates. This makes PLA$_2$ ideally suited as a sensitive probe of membrane structure; however, the mechanism of action remains unresolved; (b) PLA$_2$'s hydrolysis of lipids remains relatively unique as an example of a water-soluble enzyme requiring an insoluble substrate.

Optical evidence, using video-intensified epifluorescence microscopy of lipid monolayers in the phase transition region at the air-water interface (Grainger et al, 1989b; Grainger et al, 1990) demonstrates preferable hydrolysis of lipid solid domains, starting from points in the monolayer at the fluid-solid lipid phase boundary. After critical degrees of hydrolysis, PLA$_2$ is observed to form large, regular proteinaceous domains at the lipid interface. These enzyme domains increase in size and frequency as lipid hydrolysis progresses, retain a regular morphology throughout their growth, and are stable at surface pressures near monolayer collapse (50 mN/m) as well as at 0 mN/m. Evidence indicates that critical concentrations of fatty acid products from lipid hydrolysis are responsible for inducing enzyme domains. A further study not included in this paper has characterized mixtures of phospholipids and a polymerizable phospholipid analogue by standard monolayer techniques: isotherms, isobars, and mixing diagrams (Grainger et al, 1989a). Although these data demonstrate near ideal phase mixing behaviors in both monomeric and polymeric mixed monolayers, hydrolysis of lipid components in the film by PLA$_2$ indicates molecular structuring in the monolayer (microphase separation). Evidence indicates that PLA$_2$ hydrolysis is much more sensitive than other characterization techniques to subtle molecular level structural and compositional differences in biomembrane models.

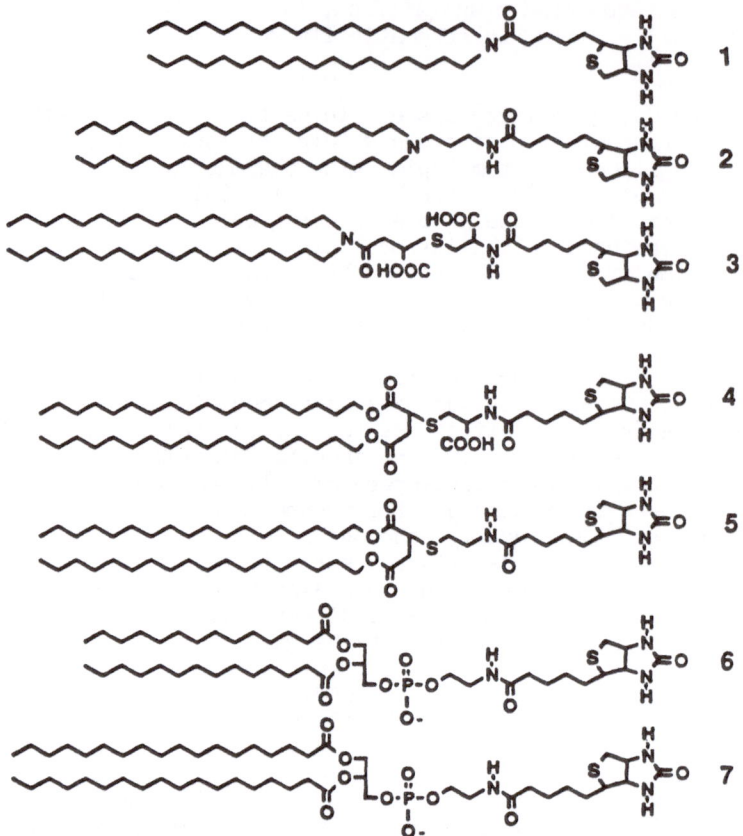

Fig. 2. Synthetic biotin lipids utilized for streptavidin binding (see Blankenburg et al, 1989).

EXPERIMENTAL

Biotin Lipid-Streptavidin Recognition and Binding in Lipid Monolayers

A series of synthetic and natural lipids having biotin covalently attached to their polar head groups (Fig. 2) have been synthesized and characterized as previously reported (Blankenburg et al, 1989). Streptavidin (Boehringer-Mannheim) was statistically labeled with fluorescein (1 and 2 labels per molecule protein) according to a standard labeling procedure (Nargessi and Smith, 1986).

Monolayer experiments. Isotherms of monolayers of biotin lipids were measured with and without various derivatives of streptavidin in the subphase using a computer-controlled monolayer film balance (Albrecht, 1989) having a Wilhelmy pressure measurement system. Lipids were spread from chloroform solutions and compressed at various rates over subphases of either distilled Millipore-filtered water, or 0.5 M NaCl (Blankenburg et al, 1989). Hysteresis experiments using streptavidin in the aqueous subphase under the lipid monolayers were performed on the same film balance at 30°C ± 0.2°C on 0.5 M NaCl subphases (Blankenburg et al, 1989). Biotin lipid monolayers were expanded into a gas-analogue state, and streptavidin in 0.5 M NaCl solution was injected into the subphase to give a total concentration of 4 μg/ml and allowed to incubate (bind) under the monolayers for two hours. The hysteresis experiment was performed by a compression to 40 mN/m, an immediate decompression to the largest molecular area possible, a subsequent 40 min incubation, and a final recompression until collapse. Compression speed was 4.5 A^2/molecule min.

Video-intensified epifluorescence microscopy of monolayers at the air-water interface. Direct visual observations of streptavidin binding to biotin lipid monolayers was achieved by using a specially designed miniaturized Langmuir film balance on the stage of an epifluorescence microscope (Meller, 1989). Fluorescence microscopes can be configured to allow direct visualization of monolayer films at the air-water interface; domains of organized molecules formed by a phase transition within the monolayer film from liquid-expanded to solid-condensed physical states are normally observed (Meller, 1988; Ahlers et al, 1990c; Losche and Mohwald, 1984). Figure 3 shows how such an instrument is designed and how a film balance may be placed on the microscope objective stage to observe monolayers at the air-water interface. Figure 4 demonstrates how this experimental configuration can be used to visualize monolayer physical states directly. Organized domains of molecules within the monolayer are formed by compressing the monolayer with a moveable barrier. A physical transition from disordered (fluid-analog) to ordered (solid-analog) state can be observed by the appearance of dark solid-ordered domains in a bright fluidized matrix. Further compression of the layer increases the size and relative surface coverage of the solid phase. Similarly, this technique has also proven valuable in investigating the interactions of labeled proteins with monolayers (Blankenburg et al, 1989; Ahlers et al, 1990a; Grainger et al, 1989b; Grainger et al, 1990). Lipid monolayers are spread at the air-water interface directly under the fluorescence microscope. Protein labeled with a fluorescent marker is introduced into the subphase under the monolayer and its interaction with the layer monitored visually over time.

Similarly to the described hysteresis procedure, biotin lipids were spread in a gas-analog state. Fluorescein-labeled streptavidin (15-20 μg in 50-100 μl 0.5 M NaCl) was injected into the subphase of this small trough (Meller, 1989) and incubated up to one hour at 30°C. Protein domains in the plane of the monolayer were visually observed using video-intensified epifluorescence microscopy of the lipid-water interface in a system previously described (Meller, 1988).

Two-dimensional protein crystals of streptavidin. Handling and Langmuir-Blodgett type transfer of large streptavidin domains formed from phospholipid-biotin lipid-mixed monolayers onto electron microscopy grids, together with protein crystallographic image analysis are detailed elsewhere (Ahlers et al, 1990b). Basically, these procedures are nearly identical to those reported for other protein crystals (Uzgiris and Kornberg, 1983; Ludwig et al, 1986).

Phospholipase A$_2$-phospholipid monolayer study. Phospholipase A$_2$ (PLA$_2$, N. naja naja), L-α-dipalmitoylphosphatidylcholine and D-α-dipalmitoyl-phosphatidylcholine (L-α-DPPC, D-α-DPPC, respectively), L-α-dipalmitoyl-phosphatidylcholine (L-α-DPPE), and L-α-dimyristoylphosphatidylcholine (L-α-DMPC) were purchased from Sigma. Dienoylphosphatidylcholine containing C-18 alkyl chains, each with butadiene groups conjugated to the glycerol backbone ester linkage in the acyl 1 and 2 positions, was synthesized by extending the methods of Patel et al (1979). A fluorescent lipid probe containing sulforhodamine in its head group was synthesized using purified N,N-dioctadecyl-1,3-propane diamine and sulforhodamine isothiocyanate (Ahlers et al, 1990c). This lipid was shown to partition preferentially into the fluid phase of monolayers.

PLA$_2$ was dissolved from the supplier's bottle in buffer and labeled (Nargessi and Smith, 1986), providing fluorescein labels on two of the estimated six lysine residues on each PLA$_2$ molecule (Grainger et al, 1989b). The resulting enzyme was indistinguishable from the unlabeled form in its

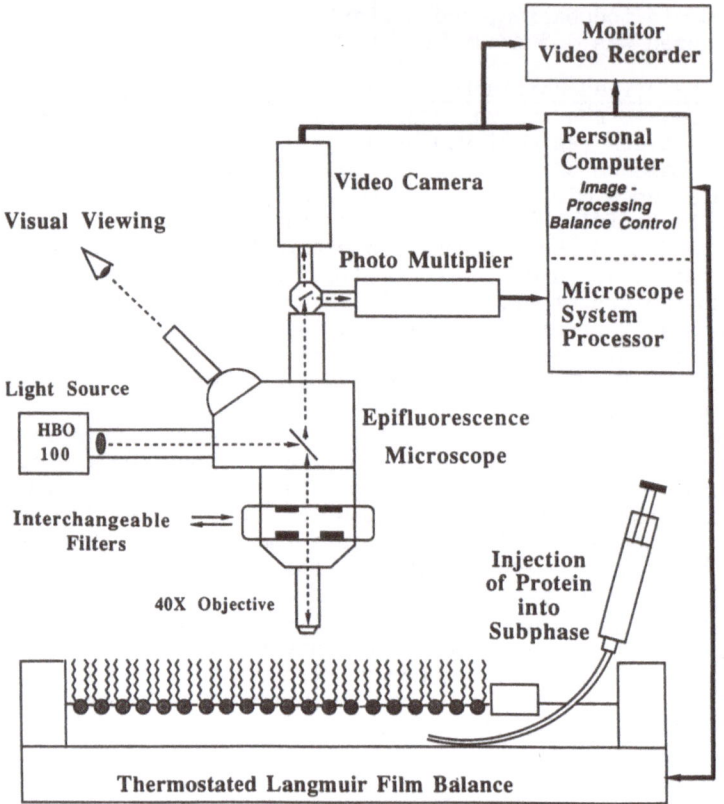

Fig. 3. Experimental configuration for epifluorescence microscopy of monolayers at the air-water interface (see Meller, 1988).

ability to hydrolyze DMPC monolayers. Enzyme solutions for interaction with monolayer experiments were made by dissolving 0.36 mg labeled PLA_2 in 26 ml buffer to make a 0.014 mg/ml solution. Aliquots of 2.1 ml were frozen in polypropylene vials at -22°C until use. For each experiment, 0.5 ml thawed PLA_2 buffer solution was removed from a vial with a glass syringe for injection under the monolayer.

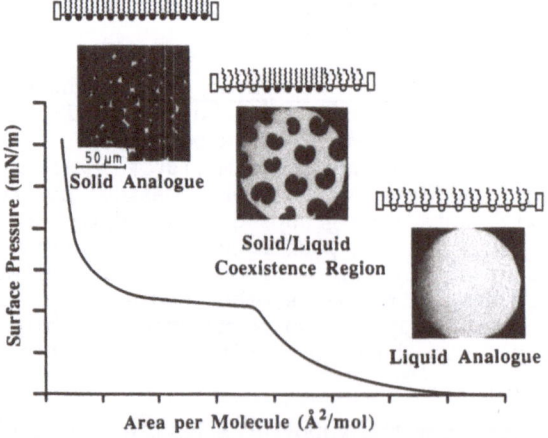

Fig. 4. Relationship between a lipid surface pressure-area diagram, its physical state, and visual observation of lipid packing using fluorescent microscopy.

Fluorescence Microscopy of Monolayers

Tris buffer (10 mM Tris, 150 mM NaCl, 5 mM $CaCl_2$, pH 8.9, chloroform washed) was prepared by dissolution of the various salts in ten times con-cencentration in acid-cleaned glass flasks. Buffer for each experiment was made by diluting the concentrated, washed stock 1:9 with pure water. Monolayers were spread over buffer at various temperatures from lipid solutions in chloroform containing 0.25-1.0 mol % of the fluorescent sulfor-hodamine lipid probe, a probe concentration dilute enough so that no changes in isotherm behavior of the main lipid component could be detected. Before spreading, the buffer surface was cleaned by suction. After spreading, the monolayer was immediately compressed by a computer-controlled barrier at a rate of 2.5 A^2/molecule per min.

At a surface pressure approximately 3-5 mN/m below the start of the phase transition of each lipid under the respective conditions, the barrier was stopped and the enzyme solution (0.5 ml) was injected from a syringe immersed from behind the barrier (see Fig. 3). After injection, the barrier was restarted and the monolayer compressed until domains of dark solid analog-phase lipids in a bright, fluid analog-lipid matrix filled the majority of the microscope field. Another strategy giving similar results involved injection of enzyme under a phospholipid monolayer in the lipid phase transition region after solid analog-lipid domains had formed. At this point, the field was observed alternately through two interchangeable fluorescent filters, corresponding respectively to the monolayer (sulforhod-amine) marker and the PLA_2 (fluorescein) marker.

Video recording of the monolayer within the mask with each filter through an SIT TV camera was initiated at various time points of film hydrolysis as described elsewhere (Grainger et al, 1989b; Grainger et al, 1990; Meller, 1988).

RESULTS AND DISCUSSION

Biotin Lipid-Streptavidin Interactions in Monolayers

The purpose of this study is a simulation of a surface recognition process by monolayer techniques. Related experimental issues including design of ligand-containing lipids for successful recognition reactions, influence of protein binding on monolayer packing, and the fate of bound protein must be clarified. In this context, it is therefore critical to distinguish between what constitutes specific recognition and processes of non-specific adsorption at interfaces. To discriminate between the two interactions, binding of active streptavidin is contrasted to binding of its inactivated form (a streptavidin-biotin complex already titrated with biotin so that no free biotin binding sites remain available). Figure 5 shows three different isotherms: a curve of the pure phospholipid-based biotin lipid 3 (broken line), another curve of the biotin lipid with active protein bound to it (solid line), and a third curve where streptavidin-$(biotin)_4$ complex was allowed to adsorb to the biotin lipid monolayer (dotted line). Indeed, specific binding can be clearly distinguished from nonspecific effects (Blankenburg et al, 1989). The binding of active streptavidin leads to a large increase in molecular area of this biotin lipid, while inter-action of inactivated streptavidin-$(biotin)_4$ with the biotin lipid changes the isotherm of the lipid very little. This suggests that inactivated streptavidin shows neither specific nor nonspecific interactions with the ligand-bound layer while active streptavidin readily and specifically binds with high affinity.

Hysteresis curves (Fig. 6) for another biotin lipid based on a cys-teine linkage (biotin lipid 7) readily distinguish between active recog-

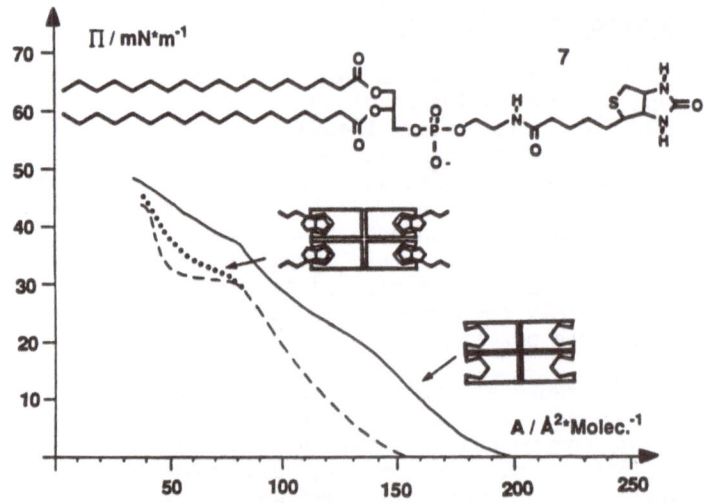

Fig. 5. Surface pressure–area isotherms of biotin lipid 3 on 0.5 M NaCl solution (pH 5.5) at 30°C: pure lipid (- - -); over subphase containing active streptavidin (————); over subphase containing inactivated streptavidin (. . .).

nition processes and adsorption tendencies. The two sets of hysteresis loops (solid and dotted lines) represent interaction with active streptavidin and inactivated streptavidin, respectively. With active streptavidin, a large hysteresis effect is observed, a large difference between expansion and recompression curves not seen in the inactivated case. Specific binding of streptavidin (solid line) causes the monolayer to relax much more slowly upon decompression, compared to nonspecific adsorption

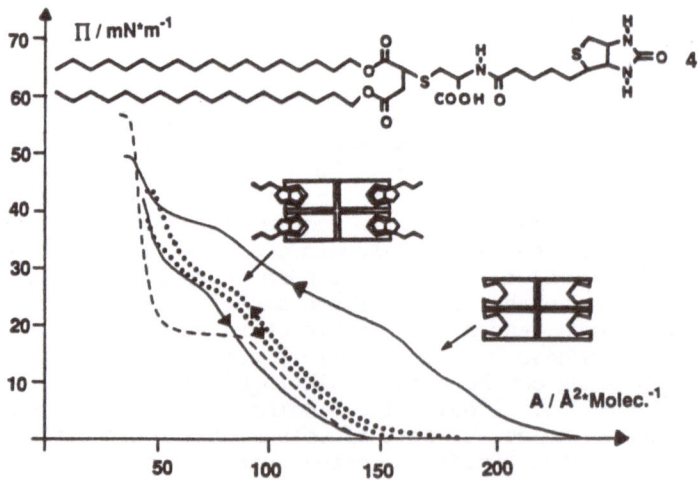

Fig. 6. Surface pressure–area isotherms of biotin lipid 7 on 0.5 M NaCl solution (pH 5.5) at 30°C: pure lipid (- - -); hysteresis over subphase containing active streptavidin, decompression and recompression (————); hysteresis over subphase containing inactivated streptavidin, decompression and recompression (. . .).

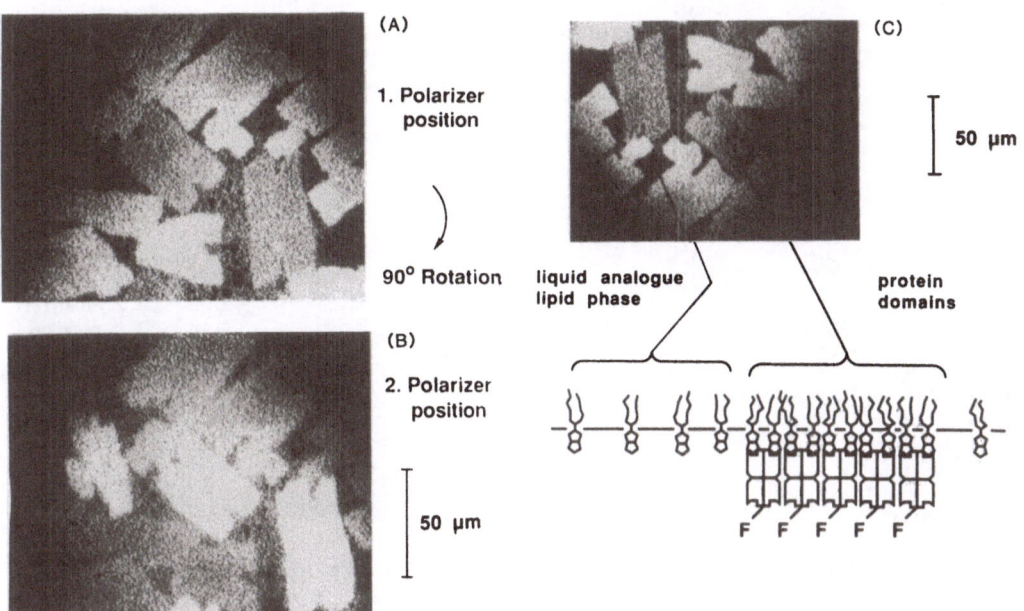

Fig. 7. Photographs from epifluorescence imaging of oriented two-
dimensional streptavidin domains bound to biotin lipid 6 at the
air–water interface (pH 5.5, 0.5 M NaCl, 30°C). Excitation light
is plane polarized and is rotated 90° in photo B, relative to
photo A. Optical anisotropy (positional and orientational) is
evident. (C) shows a schematic illustration of the domain
configuration at the interface.

(dotted line). Data from Figs. 5 and 6 strongly indicate this difference to
be due to surface recognition of active protein (Blankenburg et al, 1989).

Fluorescence Microscopy of Bound, Organized Streptavidin Domains

Epifluorescence imaging of streptavidin's interaction with biotin
lipids in monolayers gives strong evidence of protein recognition at the
interface. Figure 7 shows very typical fluorescence micrographs of streptavidin domains in the plane of the lipid monolayer, where fluorescence signal
from fluorescein-labeled streptavidin allows their direct visualization. In
this case (using monolayers of biotin lipid 6), streptavidin binds to the
lipid layer and spontaneously organizes in the plane of the interface to
generate large, two-dimensional protein assemblies. This lipid induces H-
shaped domain morphologies, exhibiting axial symmetry along two different
length axes. Very large protein domains (50–200 μm across) have been
generated and are very stable to further manipulation at the interface.
Nonspecific adsorption can easily be discounted, as no protein domains are
seen with monolayers of natural phosphatidylcholine without biotin head
groups (Blankenburg et al, 1989; Ahlers et al, 1990a).

A unique characteristic of these streptavidin domains is their optical
anisotropy. Fluorescence intensity is dependent upon the angle of polarization of the excitation beam: both photos A and B shown in Fig. 7 show the
same image changed simply by rotation of the exciting light by 90°.
Comparisons of identical domains in both photos demonstrate complete
inversion of fluorescence intensities in domains upon polarizer rotation.
Fluorescence intensity is dependent upon macroscopic domain positioning

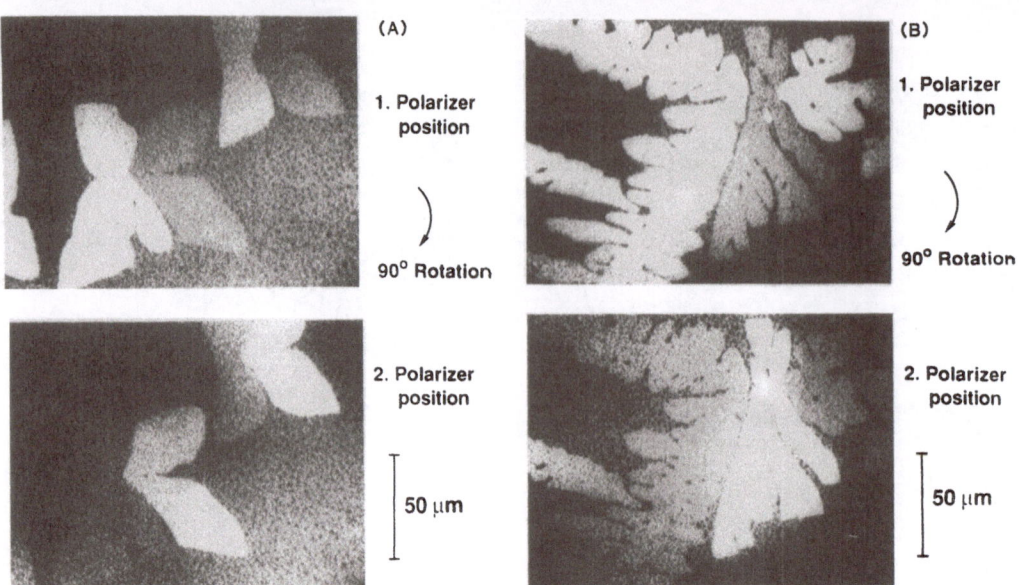

Fig. 8. Various morphologies of two-dimensional protein domains of streptavidin bound to biotin lipids at the air-water interface (pH 5.5, 0.5 M NaCl, 30°C). Polarization of excitation light is shifted 90° between each photo in A and also in B. These domain morphologies are typically observed using biotin lipids 4 and 6.

within the layer as well. Such evidence strongly indicates a regular, fixed orientation of streptavidin molecules within domains, leading to two-dimensional, oriented structures of lipid-bound protein at the interface, schematically shown in Fig. 7C. These data also suggest that the protein domains are two-dimensionally crystalline, an issue further clarified in the following section.

Many factors may contribute to streptavidin domain formation and macroscopic morphologies observed by epifluorescence microscopy. Temperature, pH, ionic strength of the subphase, as well as relative lipid and protein concentrations have been shown to affect both the kinetics and morphology of protein domain formation (unpublished evidence). Higher protein concentrations drastically accelerate domain formation time from hours to minutes. Lipid chemistry also plays an important role in influencing protein binding, orientation and organization. Indeed, Fig. 8 shows various protein domain morphologies seen with different lipids. H-shaped domains can be observed when streptavidin binds to biotin lipids 1, 2, 3, 4 and 6; spacer-less lipid 1 induces these domains only in the monolayer gas-analog state. Butterfly-shaped and dendritic or fractal structures are observed only with biotin lipids 4 and 6 (Ahlers et al, 1990a). Since lipids serve to bind and anchor protein molecules at the interfacial plane, current speculation is that the various spacer chemistries present in the various biotin lipids induce different binding geometries of streptavidin relative to the lipid monolayer plane. Lateral diffusion of lipid-bound protein causes protein assembly into larger domains. This two-dimensional organization leading to macroscopically observed protein domains of various morphologies could then be a function of initial lipid binding and recognition phenomena whose orientation is lipid- and spacer-dependent. In fact, with the spacer-less biotin lipid 1, streptavidin crystals are observed only when the biotin lipid monolayer is in a very loosely organized, gas-analog state (partial monolayer).

Only organization or compression of monolayers of biotin lipid 1 prevents protein recognition and crystallization. In contrast, spacer-containing biotin lipids induce streptavidin crystallization in densely packed monolayers at high-lateral surface pressures.

Crystalline Domains of Monolayer-Bound Streptavidin

Cooperative efforts with R.D. Kornberg and his group (Stanford University) have elegantly demonstrated by electron and optical crystallographic techniques that domains of protein seen by epifluorescence are in fact two-dimensional protein crystals (Ahlers et al, 1990a; Ahlers et al, 1990b). Figure 9 shows a high-resolution electron micrograph of the protein lattice stained with uranyl acetate. Next to it is a computer-generated cross-section of the protein molecule bound to the monolayer, created by multiple-angle transmission electron microscopy analysis of the protein lattice. By overlaying the recently published three-dimensional crystal structure of streptavidin (Hendrickson et al, 1989), the positions of the four biotin binding sites with respect to the protein and its orientation at the monolayer become known (see Fig. 10). The requirements for protein recognition, the presentation of biotin via spacers as well as the geometry of the protein molecule present at the interface, are clearly discernible. The symmetry of the protein subunits (Hendrickson et al, 1989) indicates that two binding sites at the top of the streptavidin molecule are available to bind with the biotin-lipid monolayer, while the two remaining sites located on the underside of the molecule facing the subphase are left open. The high-affinity binding and accessible nature of these biotin binding sites is ideal for high-affinity binding of further biotinylated compounds: immobilization of enzymes and antibodies, functional groups, or spacers for building multilayer proteinaceous assemblies based on an organized protein template (Ahlers et al, 1990a).

Phospholipase A$_2$ Hydrolysis of Phospholipid Monolayers and Subsequent Enzyme Domain Formation

Monolayer imaging using epifluorescence microscopy. Formation of solid lipid domains resulting from a phase transition between disordered fluid states and crystalline solid states in lipid monolayers has been modeled and visualized in a number of different fluorescent microscopy systems (Meller, 1988; Ahlers et al, 1990c; Losche and Mohwald, 1984). Enantiomerically pure, chiral phospholipids are well recognized to form solid phase domains within the lipid liquid-solid phase transition region in monolayers, easily observable with fluorescent microscopy as shown in Fig. 4 previously (Weiss and McConnell, 1984; Florsheimer and Mohwald, 1989; Gaub et al, 1986). A fluorescent lipid probe mixed in very small proportions (0.5 mol %) with the lipid of interest will readily partition into the less ordered fluid lipid phase during phase transitions in monolayers. Optical contrast arises between dark solid domains and bright fluid lipid matrix because of the fluorescent probe's exclusion from solid lipid domains. Thus, the fluid lipid phase remains greatly probe-enriched to provide contrast for imaging of the monolayer's physical states. These lipid physical states are readily seen in monolayers with epifluorescence microscopy (Meller, 1988).

Hydrolysis of phospholipid monolayers by membrane-active phospholipase A$_2$. Phospholipid monolayers containing small percentages of a sulforhodamine lipid probe were compressed over a buffer subphase at a controlled temperature such that the liquid-solid phase transition for the various lipid systems always occurred near 22 mN/m surface pressure. Dark solid lipid domains of either L-α-DPPC, D-α-DPPC, L-α-DPPE, or L-α-DMPC were visualized in bright matrices of their respective fluid states enriched with the sulforhodamine probe using the rhodamine cutoff filter. Injection of PLA$_2$ labeled with fluorescein into the subphases underneath these layers

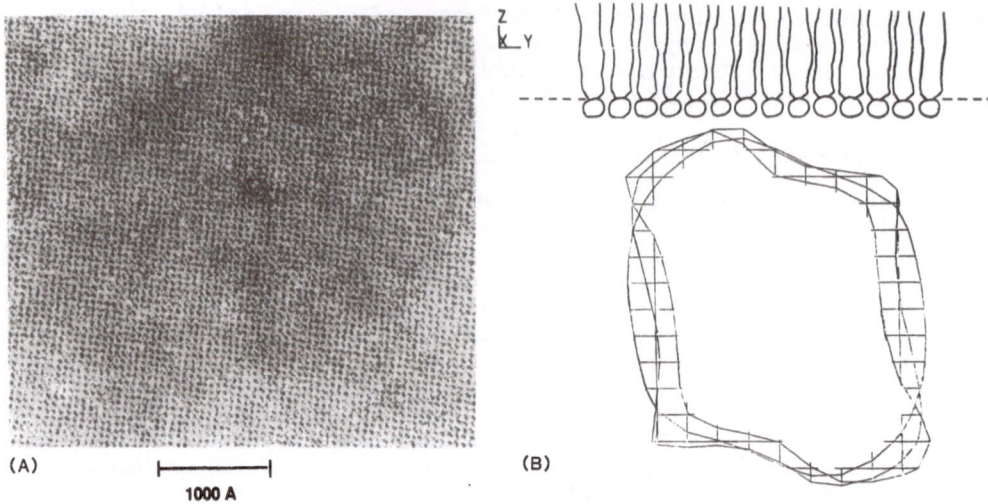

(A)

1000 A

(B)

Fig. 9.   Two-dimensional protein crystals of streptavidin induced by binding
          to biotin lipids at the air-water interface: (A) Electron micros-
          copy of a protein domain lattice; (B) A 12 A-thick slice through
          the electron density map of a bound streptavidin molecule viewed
          side-on with respect to the biotin lipid monolayer. (See Ahlers et
          al, 1990b.)

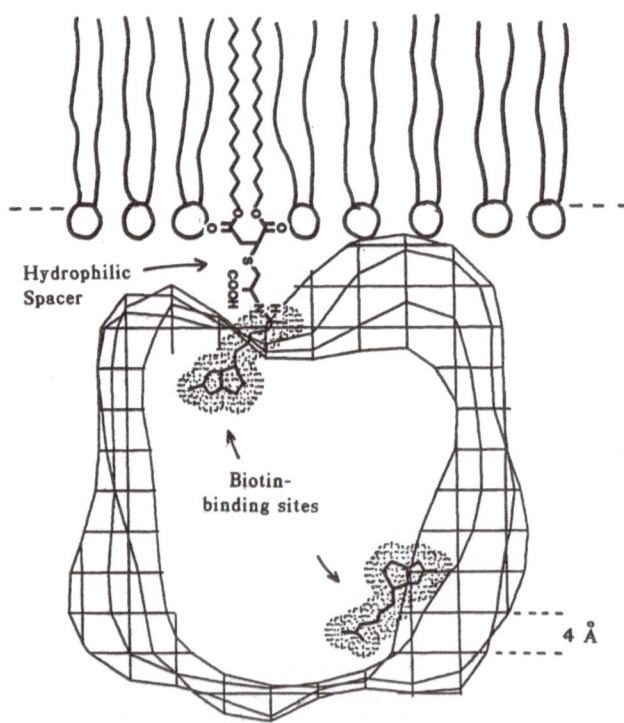

Hydrophilic
Spacer

Biotin-
binding sites

4 Å

Fig. 10.   Superposition of two biotin binding sites, determined from x-ray
           crystallography (see Weiss et al, 1984) over two-dimensional
           cross section generated from the electron image map of monolayer-
           crystallized streptavidin (Ahlers et al, 1990b). Location of
           biotin binding sites with respect to the necessary geometrical
           constraints of streptavidin's recognition and binding are evident.

provided initially a diffuse homogeneous fluorescent signal under the mono-layer seen through the fluorescein filter.

Figures 11-13 contrast the progress of enzyme recognition, binding, hydrolysis, and ultimately domain formation in monolayers of the various phospholipids (Grainger et al, 1989b; Grainger et al, 1990). Figure 11 displays the time-dependent hydrolysis of L-α-DPPC monolayers by PLA_2. Figure 11A shows the monolayer as seen through the rhodamine filter, while Fig. 11B shows the same image through the fluorescein filter where signal is generated by fluorescently labeled enzyme under the monolayer immediately after injection. Starting from timepoint zero (immediately after enzyme introduction; Fig. 11A, rhodamine filter, and Fig. 11B, corresponding fluorescein filter), the enzyme binds to phospholipids in the monolayer and starts to hydrolyze. These enzyme recognition sites are often the clefts, discontinuities, and indentations in the sides and edges of the L-α-DPPC solid lipid domains. PLA_2 hydrolysis as observed through the rhodamine filter proceeds directionally from these initial binding sites on the inter-facial boundary between solid and liquid lipid phases into the interior of the solid lipid domains (Figs. 11C, 11E). Observation through the fluores-cein filter (Fig. 11D) shows that, at certain locations in the partially hydrolyzed monolayer, bright domains of enzyme have formed. These domains of labeled enzyme increase in size as hydrolysis continues with time (Fig. 11F). At later timepoints (60 min, Fig. 11G), enzyme hydrolysis has destroyed nearly all solid lipid domains. Moreover, the small enzyme aggregates seen at earlier timepoints (Figs. 11D, 11F) have grown to form large, regular enzyme domains of consistent morphology (bright domains seen with fluorescein filter, Fig. 11H) (Grainger et al, 1989b).

This pattern of hydrolytic behavior, the binding, hydrolysis and PLA_2 domain formation, is witnessed consistently in phospholipid monolayers as long as hydrolysis occurs. Figure 12 shows an analogous series of time-dependent photos in the hydrolytic sequence for L-α-DMPC monolayers in this lipid's fluid-solid phase transition. Lipid solid domains (Fig. 12A) are smaller and show different morphology than those of L-α-DPPC. Yet, hydro-lysis of the layer by PLA_2 demonstrates the same enzyme domain formation behavior as seen in the L-α-DPPC case (Grainger et al, 1990). Through the rhodamine filter (Fig. 12, left column), lipid domains are attacked by the enzyme on their outer edges (Fig. 12B), resulting in small, bright enzyme aggregates located where hydrolysis has penetrated into the solid lipid islands (see arrow, fluorescein filter, Fig. 12C). At longer hydrolysis times, large circular areas have been hydrolyzed out of the lipid monolayer (Figs. 12D, 12F) and large enzyme domains (corresponding bright regions, Figs. 12E, 12G) can be seen through the fluorescein filter (right column, Fig. 12). Finally, the enzyme hydrolyzes most of the lipid domains away (Fig. 12H), leaving large, regular enzyme domains of identical morphology to those seen with L-α-DPPC in the monolayer (compare Figs. 12H, 12I).

An example of PLA_2 hydrolysis of L-α-DPPE monolayers is shown in Fig. 13 (Grainger et al, 1990). The first photo (Fig. 13A) shows the typical, denritic types of solid lipid domains for this lipid in its phase trans-ition region. Figure 13B shows enzyme hydrolysis of these solid domains upon introduction of PLA_2 (as seen through the rhodamine filter). The arrow points to an enzyme domain, formed within a solid lipid domain remnant. At later time (Fig. 13C), the lipid domains are nearly totally destroyed and enzyme domains (compare with corresponding fluorescein filter view in Fig. 13D) are large and numerous (Grainger et al, 1990). This behavior is consistent with the other cases of phosphatidylcholines and, despite the difference in head group chemistries, seems to indicate that lipid hydro-lysis itself leads to induction of enzyme domains in the layer.

Moreover, experiments using non-hydrolyzable lipids, D-α-DPPC and the dienoylphosphatidylcholine polymerizable lipid analog, demonstrate how

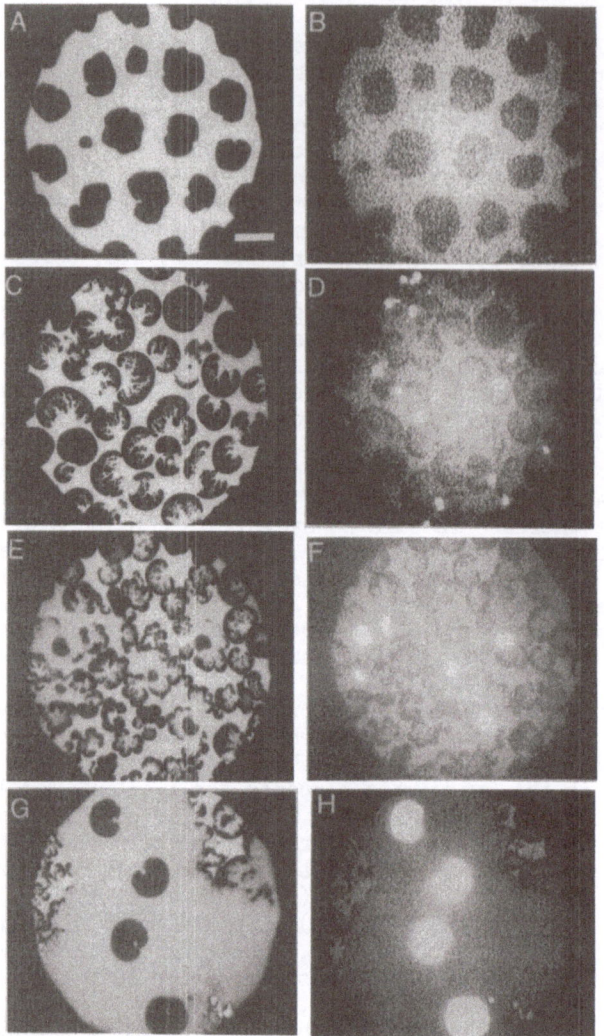

Fig. 11. Real-time epifluorescent observation of the hydrolytic action of
fluorescein-labeled PLA$_2$ on monolayers of L-α-DPPC in the phase-
transition region, constant surface pressure of 22 mN/m, buffer
subphase, temperature = 30°C: (A) Time = 0 (immediately after
PLA$_2$ injection into subphase, rhodamine filter); (B) same view as
in (A) seen through fluorescein filter (diffuse enzyme fluorescent signal from subphase); (C) Hydrolysis of monolayer at time =
25 min (rhodamine filter). Domains of solid lipid have been substantially hydrolyzed by PLA$_2$. (D) Same image as in (C) seen
through fluorescein filter. Small bright enzyme domains have
been formed in the layer. (E) Further lipid domain hydrolysis
at time = 40 min (rhodamine filter). (F) Same image as in (E)
seen in fluorescein filter. Enzyme domains are larger and their
morphology is well developed and regular. (G) Significant enzyme
hydrolysis has destroyed most solid domains of lipid at time = 60
min (rhodamine filter). Large, kidney-shaped domains are enzyme
(see (H)). (H) Same image as in (G) seen in fluorescein filter.
Large enzyme domains have formed having a regular shape and size.
Scale bar in (A) is 20 μm.

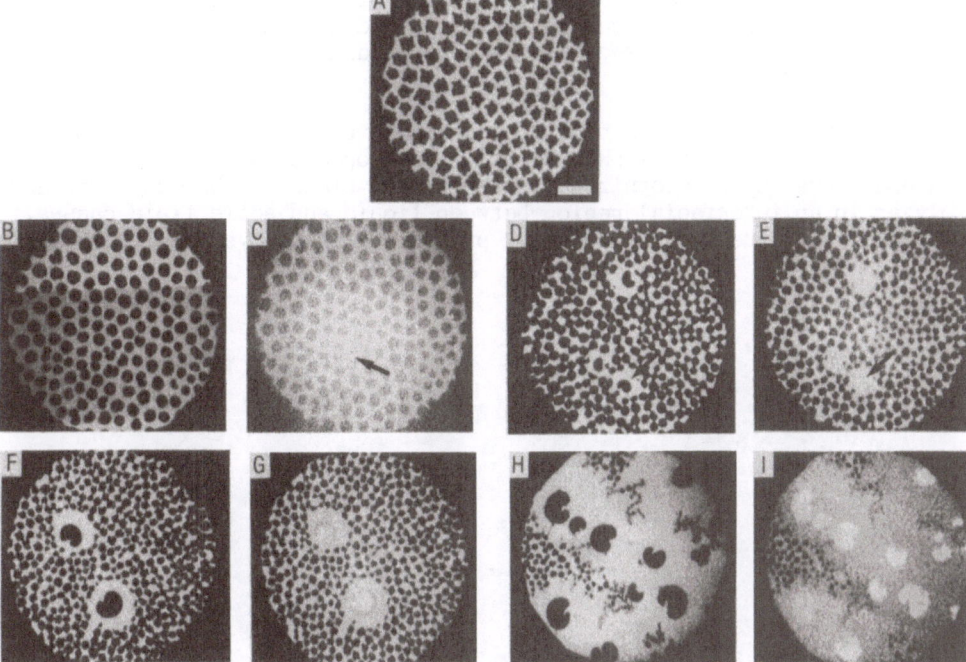

Fig. 12.   Dual epifluorescence imaging of L-α-DMPC monolayer phase transition region and subsequent hydrolysis by injection of phospo-
lipase A_2 into the aqueous subphase under the monolayer. (A) L-α-
DMPC monolayer seen through the sulforhodamine filter after compression into phase transition region as described in text, and
(B) same monolayer as in (A) after 30 min annealing time and after
enzyme addition. (C) Monolayer as in (B) seen through the fluorescein filter after enzyme addition. Note small bright points
(arrow) indicating enzyme attachment to lipid monolayer. (D) Same
monolayer at 25 min hydrolysis time seen through the sulforhodamine filter. Hydrolyzed areas are clearly seen along with
accompanying enzyme domain formation. (E) Same image as in (D)
but seen through the fluorescein filter. Bright areas (arrow)
are enzyme domains. (F) Same monolayer at 40 min hydrolysis time
imaged through the rhodamine filter. Monolayer hydrolysis is
substantial and enzyme domains have grown in size but maintain
original morphology. (G) Same image as in (F) but seen through
the fluorescein filter. Bright areas are enzyme domains.
(H) Same monolayer after 55 min hydrolysis time imaged through
rhodamine filter. The black solid phase DMPC domains are nearly
completely destroyed and numerous domains of enzyme (large
kidney-shaped domains in the bright fluid monolayer phase) have
assembled at the interface. (I) Corresponding image to (H) seen
through the fluorescein filter. Enzyme domains are readily
identified as fluorescing, bright areas. Scale bar in (A) is 20
μm. Temperature = 10°C, surface pressure = 22 mN/m.

71

critical hydrolysis is. Although monolayers of these compounds demonstrate phase transitions and lipid solid domain formation, they remain inert to PLA hydrolysis. Furthermore, fluorescence of PLA_2 at the interface is diminished, indicating little recognition or binding. Finally, no aggregate or domains of enzyme are ever observed to form within these layers, indicating that a sequence of recognition, binding and hydrolytic action are necessary precursors to the mechanism of enzyme domain formation (Grainger et al, 1989b; 1990).

A mechanism for the formation of organized, two-dimensional enzyme domains within monolayers of phospholipids is proposed in Fig. 14. Active enzyme under the layer recognizes its substrate, binds to the monolayer and hydrolyzes in an interfacial region between liquid and solid lipid phases. After a critical extent of hydrolysis, products of hydrolysis - lysolipids

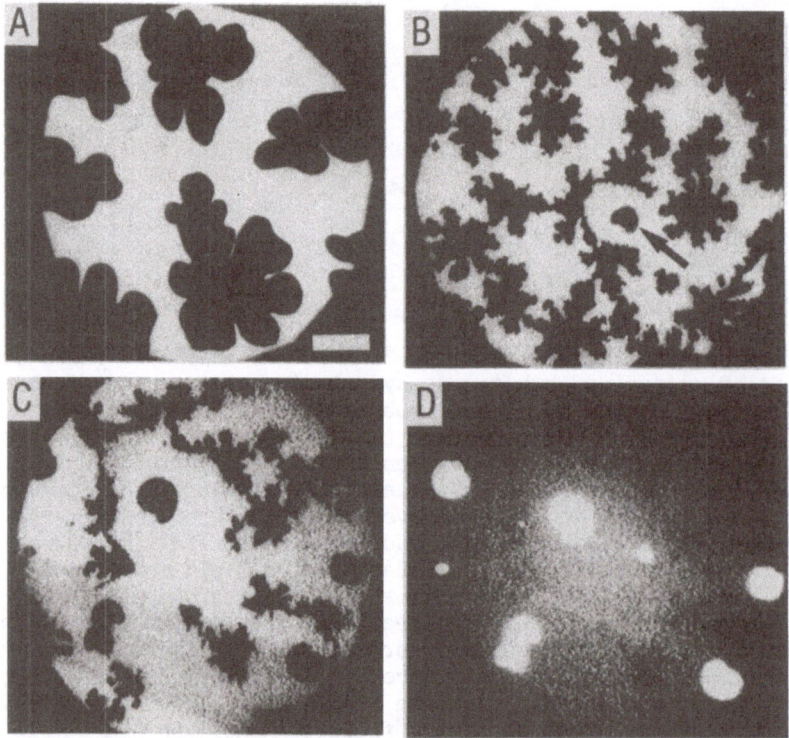

Fig. 13. Epifluorescence imaging of L-α-DPPE monolayer phase transition region and subsequent hydrolysis by injection of phospholipase A_2 into the aqueous subphase under the monolayer. Identical experimental conditions as in Fig. 11 except that the temperature = 37°C. (A) L-α-DPPE monolayer in the phase transition region before enzyme addition showing typical dendritic type of domains for this solid phase. (B) through (D) show progress of enzyme hydrolysis at time = : (B) 30 min hydrolysis time, imaged through the sulforhodamine filter. Note frayed, spiked lipid solid domain edges and formation of small enzyme domain (arrow); (C) 45 min hydrolysis time. Lipid solid domains are nearly completely destroyed and many enzyme domains can be seen in the layer; (D) identical view as in (C) except seen through the fluorescein filter. Bright areas are corresponding enzyme domains. Scale bar in (A) is 20 μm.

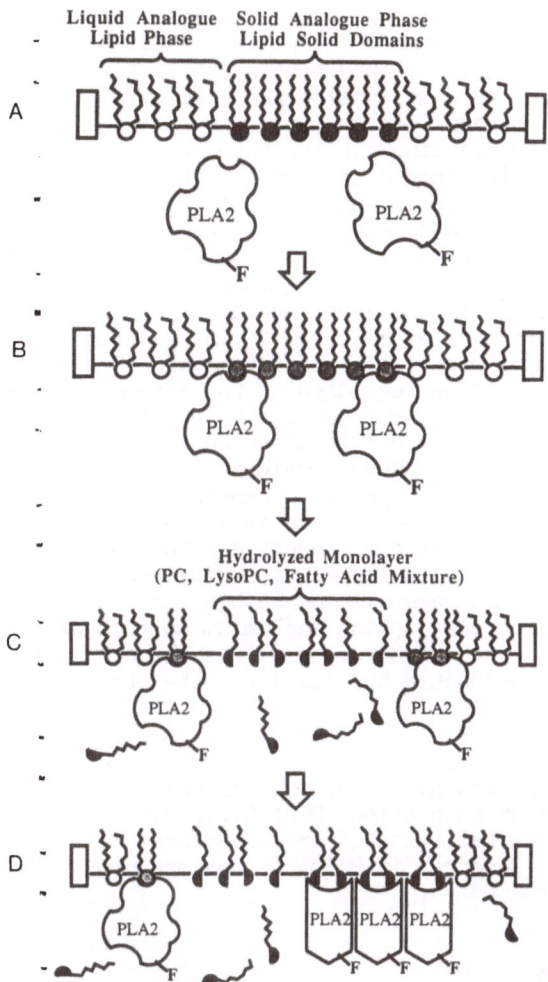

Fig. 14. Proposed mechanism for recognition, binding, hydrolysis and subsequent phospholipase A_2 domain formation prompted by critical
concentrations of hydrolytic end-products mixed with substrate
lipids in phospholipid monolayers: (A) Injection of phospholipase
A_2 into aqueous subphase under lipid monolayer; (B) recognition
and binding of phospholipase A_2 to lipid interface; (C) hydrolysis of monolayer by phospholipase A_2 with subsequent build-up
of hydrolytic products (fatty acid and lysolipid) within the
layer; (D) organization of bound phospholipase A_2 into protein
domains at the lipid-water interface prompted by critical concentrations and phase separation of hydrolytic end-products in the
monolayer.

and fatty acids – build up in localized regions of the layer. Phase separation of these products from pure lipid is proposed to occur, leading to
areas of increased charge density in the case of fatty acids. In fact, in
vesicle systems, such fatty acid phase separation and charge densities have
already been shown (Yu et al, 1989). Enzyme may then be prompted to bind
and build domains in these areas of localized charge, leading to the phenomena of enzyme domain formation witnessed in DPPC, DMPC and DPPE monolayers
(Grainger et al, 1990).

Enzyme domain stability. In order to assess the structural integrity
and stability of the two-dimensional protein domains resulting from monolayer hydrolysis, hydrolyzed lipid monolayers containing large enzyme
domains were expanded to a gas-analog phase (0 mN/m surface pressure) as
well as compressed to near film collapse (50 mN/m surface pressure (Grainger
et al, 1990). Two interesting results were observed. As monolayers were
expanded, enzyme domains tended to aggregate together in groups, but in no
case was there any evidence to suggest that surface tension forces had
fractured or pulled enzyme domains apart. Even in the gas-analog phase
(figure not shown), intact enzyme domains could be observed at the air-water
interface. Moreover, at extremely high surface pressures (up to 50 mN/m),
enzyme domains remained in the layer and did not alter their structure in
any apparent way (Grainger et al, 1990). Interestingly, unhydrolyzed lipid
remaining in the layer recrystallized into solid domains again at high surface pressures, often using the edges of the enzyme domains as nucleation
points (figure not shown). Some domains were completely encircled by a
crystallized lipid border. This leads to an interesting question as to
whether enzyme molecules in these domain structures remain active or not.
Although solid lipid domains appear to remain intact along the interfaces of
enzyme domains in the monolayer, there is yet no concrete evidence that
enzyme in domains is inhibited or inactive. The fact that these proteinaceous assemblies are so morphologically homogeneous, regardless of the
lipid chemistry used, and are also so stable against mechanical forces
suggests that some regular packing mechanism must control their assembly at
the interface. That such structuring allows the enzyme to remain hydrolytically active is questionable but remains to be proven.

Acknowledgements

D.W.G. was supported by a Postdoctoral Research Fellowship from the
Alexander von Humboldt Foundation, West Germany. C.S. was supported by a
Postdoctoral Research Fellowship from the Natural Sciences and Engineering
Council of Canada. Research in Mainz is assisted by a BMFT Grant for
Ultrathin Films (03M4008F1).

REFERENCES

Ahlers, M., Blankenburg, R., Grainger, D.W., Meller, P., Ringsdorf, H. and
    Salesse, C., 1990a, Specific recognition and formation of two-
    dimensional streptavidin domains in monolayers: applications to
    molecular devices, Thin Solid Films, 180:93.
Ahlers, M., Blankenburg, R., Darst, S.A., Kornberg, R.D., Kubaleck, E.W.,
    Ribi, H.O. and Ringsdorf, H., 1990b, Two dimensional crystallization
    of streptavidin induced by specific binding to biotin-lipid mono-
    layers, Biochemistry, submitted.
Ahlers, M., Grainger, D.W., Ringsdorf, H. and Salesse, C., 1990c, New
    fluorescent lipids as membrane probes: synthesis and character-
    zation, Chem.Phys.Lipids, submitted.
Albrecht, O., 1989, The construction of a microprocessor-controlled film
    balance for precision measurement of isotherms and isobars,
    Thin Solid Films, 99:227.
Blankenburg, R., Meller, P., Ringsdorf, H. and Salesse, C., 1989,
    Interaction between biotin lipids and streptavidin in monolayers:
    formation of oriented two-dimensional protein domains induced by
    surface recognition, Biochemistry, 28:8214.
Bundgaard, H., Hansen, A.B. and Kofod, H., eds., 1981, "Optimization of Drug
    Delivery", Munksgaard, Copenhagen.
Chen, P., Pearce, D. and Verkman, A.S., 1988, Membrane water and solute
    permeability determined quantitatively by self-quenching of an
    entrapped fluorophore, Biochemistry, 27:5713.

Dhathathreyan, A. and Mobius, D., 1988, Local anesthetics-phospholipid interaction. A study of dibucaine binding to lipid monolayers, Coll.Surf., 33:43.

Dhathathreyan, A., Baumann, U., Muller, A. and Mobius, D., 1988, Characterization of complex gramicidin monolayers by light reflection and Fourier transform infrared spectroscopy, Biochim.Biophys.Acta, 944:265.

Fendler, K., Grell, E. and Bamberg, E., 1987, Kinetics of pump currents generated by the sodium-potassium ATPase, FEBS Lett., 224:83.

Florsheimer, M. and Mohwald, H., 1989, Development of equilibrium domain shapes in phospholipid monolayers, Chem.Phys.Lipids, 49:231.

Gaub, H., Buschl, R., Ringsdorf, H. and Sackmann, E., 1985, Phase transitions, lateral phase separation and microstructure of model membranes composed of a polymerizable two-chain lipid and dimyristoylphosphatidylcholine, Chem.Phys.Lipids, 37:19.

Gaub, H.E., Moy, V.T. and McConnell, H.M., 1986, Reversible formation of plastic two-dimensional lipid crystals, J.Phys.Chem., 90:1721.

Grainger, D.W., Reichert, A., Ringsdorf, H., Salesse, C., 1989a, Hydrolytic action of phospholipase A$_2$ in monolayers in the phase transition region: direct observation of enzyme domain formation using fluorescence microscopy, Biochim.Biophys.Acta, in press.

Grainger, D.W., Reichert, A., Ringsdorf, H. and Salesse, C., 1989b, An enzyme caught in action: direct imaging of hydrolytic function and domain formation of phospholipase A$_2$ in phosphatidylcholine monolayers, FEBS Lett., 252:73.

Grainger, D.W., Reichert, A., Ringsdorf, H., Salesse, C., Davies, D. and Lloyd, J.B., 1990, Mixed monolayers of natural and polymeric phospholipids: structural characterization by physical and enzymatic methods, Biochim.Biophys.Acta, in press.

Gregoriadis, G., Allison, A., eds., 1980, "Liposomes in Biological Systems", J. Wiley, Chichester.

Hashimoto, K., Loader, J.E., Knight, M.S. and Kinsky, S.C., 1985, Inhibition of cell proliferation and dihydrofolate reductase by liposomes containing methotrexate-dimyristoylphosphatidylethanolamine derivatives and by the glycerophosphorylethanolamine analogs, Biochim.Biophys.Acta, 816:169.

Heckl, W.M., Zaba, B.N. and Mohwald, H., 1987, Interactions of cytochromes b5 and c with phospholipid monolayers, Biochim.Biophys.Acta, 903:166.

Hendrickson, W.A., Pahler, A., Smith, J.L., Satow, Y., Merritt, E.A. and Phizackerley, R.P., 1989, Crystal structure of core streptavidin determined from multiwavelength anomalous diffraction of synchrotron radiation, Proc.Natl.Acad.Sci.USA, 86:2190.

Ibdah, J.A. and Phillips, M.C., 1988, Effects of lipid composition and packing on the adsorption of apolipoprotein A-1 to lipid monolayers, Biochemistry, 27:7155.

Losche, M. and Mohwald, H., 1984, Fluorescence microscope to observe dynamical processes in monomolecular layers at the air/water interface, Rev.Sci.Instrum., 55:1968.

Ludwig, D.S., Ribi, H.O., Schoolnik, G.K. and Kornberg, R.D.,1986, Two-dimensional crystals of cholera toxin B-subunit-receptor complexes: projected structure at 17-A resolution, Proc.Natl.Acad.Sci.USA, 83:8585.

Matsuhita, T., Fyu, E.K., Hong, C.I. and MacCoss, M., 1981, Phospholipid derivatives of nucleoside analogs as prodrugs with enhanced catabolic stability, Cancer Res., 41:2707.

Mayer, L.D., Wong, K.F., Menon, K., Chang, C., Harrigan, P.R., Cullis, P.R., 1988, Influence of ion gradients on the transbilayer distribution of dibucaine in large unilamellar vesicles, Biochemistry, 27:2053.

Meller, P., 1988, Computer-assisted video microscopy for the investigation of monolayers on liquid and solid substrates, Rev.Sci.Instrum., 59:2225.

Meller, P., 1989, Microspectroscopy on single domains of phase-separated monolayers, J.Micros.(Oxford), 156:241.

Mohwald, H., 1988, Lateral molecular organization and order in monomolecular layers, J.Mol.Elect., 4:47.

Nargessi, R.D. and Smith, D.S., 1986, Fluorometric assays for avidin and biotin, Methods Enzymol., 122:67.

Ohtoyo, T., Shimagaki, M., Otada, K., Kimura, S. and Imanashi, Y., 1988, Change in membrane fluidity induced by lectin-mediated phase separation of the membrane and agglutination of phospholipid vesicles containing glycopeptides, Biochemistry, 27:6458.

Patel, K.M., Morisett, J.D. and Sparrow, J.T., 1979, A convenient synthesis of phosphatidylcholines: acylation of sn-glycero-3-phosphocholine with fatty acid anhydride and 4-pyrrolidinopyridine, J.Lipid Res., 20:674.

Ringsdorf, H., Schlarb, B. and Venzmer, J., 1988, Molecular architecture and function of polymeric oriented systems: models for the study of organisation, surface recognition and dynamics of biomembranes, Angew.Chem.Int.Ed.Engl., 27:113.

Rosemeyer, H., Ahlers, M., Schmidt, B. and Seela, F., 1985, A nucleolipid with antiviral acycloguanosine as a head group-synthesis and liposome formation, Angew.Chem.Int.Ed.Engl., 24:501.

Sackmann, E., Duwe, H.P., Zeman, K. and Zilker, A., 1986, Elasticity, structure and dynamics of cell plasma membrane and biological functions, in: "Structure and Dynamics of Nucleic Acids, Proteins and Membranes", E. Clement, S. Chin, eds., Plenum, New York.

Thompson, N.L., Palmer, A.G. III, Wright, L.L. and Scarborough, P.E., 1988, Fluorescence techniques for supported planar model membranes, Comm.Mol.Cell.Biophys., 5:109.

Uzgiris, E.E. and Kornberg, R.D., 1983, Two-dimensional crystallization technique for imaging macromolecules, with application to antigen-antibody-complement complexes, Nature (London), 301:125.

Venema, F.R. and Weringa, W.D., 1988, The interaction of phospholipid vesicles with some anti-inflammatory agents, J.Colloid Interface Sci., 125:484.

Weiss, R.M. and McConnell, H.M., 1984, Two-dimensional chiral crystals of phospholipid, Nature, 310:47.

Weiss, R.M., Tamm, L.K., McConnell, H.M., 1984, Periodic structures in lipid monolayer phase transitions, Proc.Natl.Acad.Sci.USA, 81:3249.

Yu, B-Z., Kozubek, Z. and Jain, M.K., 1989, Binding of phospholipase A_2 to zwitterionic bilayers is promoted by lateral segregation of anionic amphiphiles, Biochim.Biophys.Acta, 980:23.

ENDOCYTOSIS OF CELL SURFACE MOLECULES AND INTRACELLULAR DELIVERY OF MATERIALS ENCAPSULATED IN TARGETED LIPOSOMES

Patrick Machy and Lee Leserman

Centre d'Immunologie INSERM–CNRS de Marseille–Luminy
Case 906, 13288 Marseille Cedex 9, France

INTRODUCTION

The principle benefits for pharmacologic purposes of liposomes include the possibility of transporting highly concentrated reagents entrapped in the liposomes' aqueous spaces or lipid bilayer and their protection against enzymatic degradation. In addition, certain molecules which may have little or no capacity to enter into cells because of their size or charge may be delivered in biologically active form inside cells by virtue of their association with liposomes. This capacity is enhanced when liposomes are targeted to the cell surface by the use of specific ligands such as monoclonal antibodies, which may additionally offer the possibility to concentrate material at the cell surface and to target sub-populations in a heterogeneous mixture of cells (for reviews see Leserman and Machy, 1987; Ostro, 1987).

The delivery of targeted liposome contents depends on several different parameters. These include: the size of the liposomes, the nature of the cell surface target molecule, and the cell type expressing it. Of these, the key element governing liposome entry is the endocytosis of target cell surface molecules. The area of our greatest ignorance relates to the problem of how encapsulated or lipid-associated molecules escape from liposomes and from endocytic vesicles in intact form to reach their site of action either in the cytoplasm or in the nucleus.

We will not discuss here the significance for cell function of the differential endocytosis of the cell surface molecules, which depends on the molecules in question and the cell expressing them, but we would like rather to describe what we have observed in terms of the endocytosis of the liposomes and the target molecules, and the behavior of this complex inside cells. From these observations we will conclude that understanding of the endocytic process, of the behavior of target molecules, of the liposomes and of the encapsulated molecules is essential in order to optimize delivery.

Monoclonal antibodies or other ligands such as protein A, which binds to the Fc region of antibodies, can be reproducibly coupled to the surface of the liposomes without changing or altering the specificity of the ligands for their target molecules and without enhancing leakage of the liposome contents (for review see Leserman and Machy, 1987). Ligand-bearing liposomes can bind to the cell surface with a specificity corresponding to that

Table 1. Effect of the Size of Liposomes on the Delivery Potential of Liposome Contents

Cell type	Size of liposomes in nm.		
	60–80	200	400
Fibroblasts (L cells)		Binding	
	+	++	++
		Drug effect	
	+++	++	+
T lymphocytes (spleen blasts and tumor cells)		Binding	
	+	++	++
		Drug effect	
	+++	–	–

Cells were incubated with an anti–H–2K monoclonal antibody and protein A-coupled liposomes of different sizes containing carboxyfluorescein and methotrexate. The same lipid was used to prepare all liposomes. The binding of liposomes was evaluated by spectrofluorometry and the drug effect measured by inhibition of incorporation of tritiated deoxyuridine (Machy and Leserman, 1983).

of the targeting ligand (Leserman et al, 1980; Barbet et al, 1981). This has been shown by spectrofluorometry (Truneh et al, 1983a) and cytofluorography (Truneh et al, 1987) using fluorochrome-containing targeted liposomes. Because the specificity is not altered it was also possible to show that the number of liposomes that bind to the cell is proportional to the number of molecules which are recognized by the coupled antibody (Machy et al, 1982a). This is true for small (50 to 100 nm in diameter) or large (200 nm in diameter and larger) liposomes, with monoclonal antibodies or with protein A (Machy and Leserman, 1983). In addition, because of their property of encapsulating hundreds or thousands of fluorescent molecules (i.e. carboxyfluorescein), targeted liposomes permit visualization of cell surface molecules expressed at very low levels (as few as 500 copies), which may be undetectable by other classical reagents (i.e. fluorescein-coupled antibodies) (Truneh and Machy, 1987).

There are three different parameters to be considered concerning the entry of targeted liposomes inside cells. The first refers to the size of the liposomes. As summarized in Table 1, the larger the liposomes the less efficiently they are taken up by cells and consequently, the less efficiently they deliver their contents. This has been shown, for example, by using antibodies directed against the murine major histocompatibility complex (MHC)-encoded class I H–2K molecules expressed by T lymphoblasts and liposomes containing the drug methotrexate (MTX), which inhibits cell proliferation. While liposomes of different sizes bind to the cells, liposomes of 60–80 nm, but not liposomes of 200 or 400 nm in diameter permit drug delivery and inhibition of cell proliferation. This is particularly evident with lymphocytes which are known to have poor pinocytic and phagocytic capabilities (Goldmacher et al, 1986) as compared to fibroblasts. Because fibroblasts are more phagocytic, they are more susceptible to drug effects with large liposomes than lymphocytes, even though small liposomes are always superior. These results indicate that there is little or no fusion of the liposomes bound to the plasma membrane, since, if this were the case, large liposomes, which have more drug entrapped within them than small liposomes,

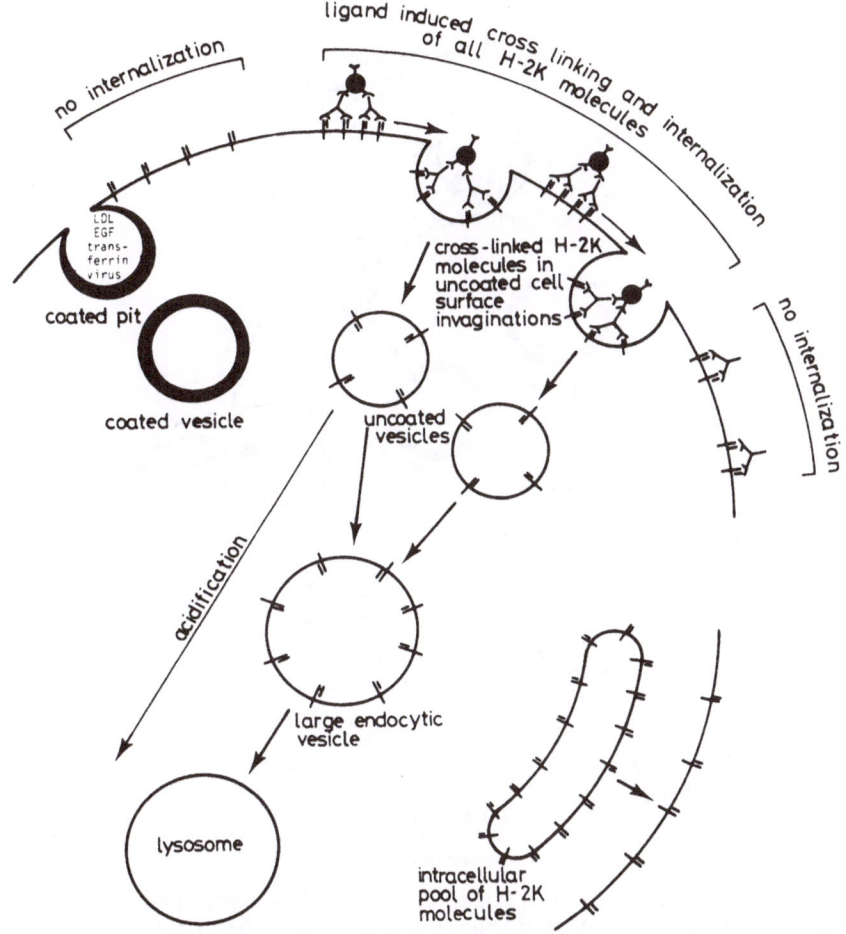

Fig. 1. Behavior of MHC class I molecules in fibroblasts (L cells).
|ı Class I H-2K molecules, ⋎ Anti-H-2K antibody, ⊸• Cross
linker (protein A-bearing liposomes).

would be more effective (Machy and Leserman, 1983). That endocytosis
governs the entry of liposomes inside the cell can be concluded from the
above results, indicating that liposomes larger than the size of endocytic
vesicles (100-150 nm), will not be internalized. The assumption of endocytosis is also in agreement with the inhibition of liposome internalization
observed when cells were treated with cytochalasins, which disrupt cyto-
skeletal elements necessary for endocytosis (Truneh et al, 1984).

The second point to be emphasized is that if one looks at the endocytic
pathway of a cell surface determinant, it is clear that the behavior of a
bound liposome can be explained by the behavior of the target molecule. The
same molecule expressed by different cell types may behave differently with
respect to endocytosis, and thus may give different results in terms of
delivery of liposome contents. The example chosen is again the murine MHC
class I molecules H-2K and H-2D, which are expressed by all nucleated
mammalian cells. When these molecules are expressed by the fibroblast-like
L cells, the liposomes are efficiently internalized and deliver their encapsulated MTX, resulting in inhibition of cell proliferation (Truneh et al,
1983a). If we look at the behavior of the liposomes and of the target class

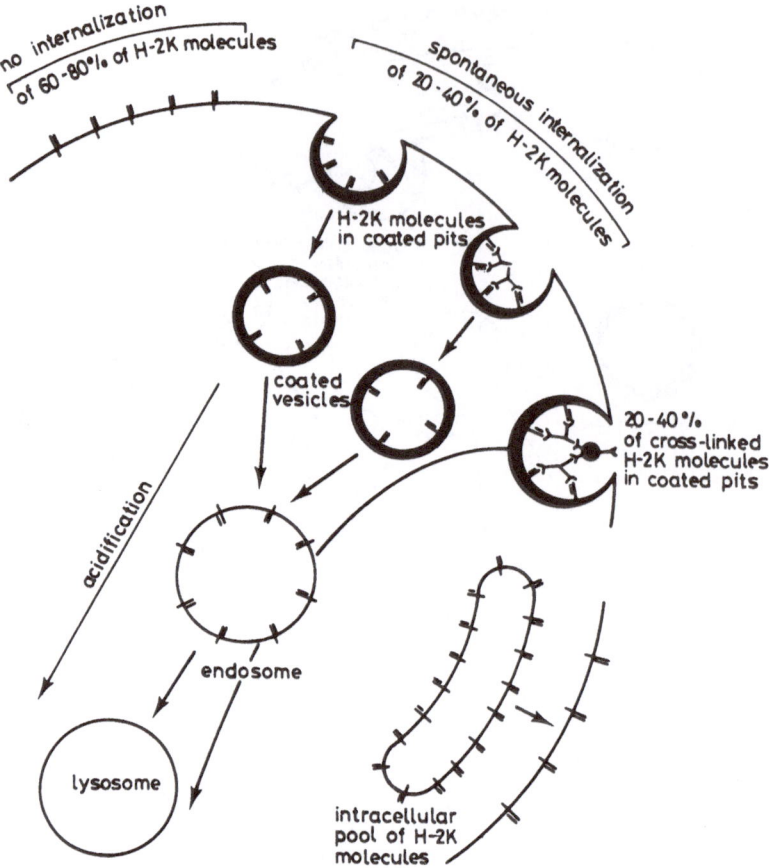

Fig. 2. Behavior of MHC class I molecules in T lymphocytes. |\ Class I
H-2K molecules, Ⴤ Anti-H-2K antibody, ⋙ Cross linker
(protein A-bearing liposomes).

I molecules by biochemical techniques and by electron microscopy, it appears
that the behavior of the class I molecules differs depending on whether or
not liposomes are bound. Normally, class I molecules are not spontaneously
internalized by fibroblasts, and the half life of the molecules is greater
than twenty hours (Huet et al, 1980; Machy et al, 1987a). However, when
cross-linked by multivalent ligands, such as liposomes which have several
antibody or protein A molecules on their surface (Barbet et al, 1981), 100%
of class I molecules are internalized with a half life of about one hour and
degraded in about five hours (Machy et al, 1987a). Thus, with fibroblasts,
the target class I molecules are internalized only after cross linking. In
addition, the induced internalization involved non-clathrin coated vesicles,
although fibroblasts have many coated pits. This is summarized in Fig. 1.
It is probably because smooth endocytic vesicles are implicated in the
internalization of class I molecules that fibroblasts are able to internalize large liposomes, even though less efficiently compared to small ones
(see above). For the same molecules, if one examines lymphocytes using the
same liposomes and the same antibodies, it has been found that T lymphocytes
are susceptible to the encapsulated drug (Machy et al, 1982a, b). In this
case the target MHC molecules are spontaneously internalized with a half
life of one hour, and to a proportion which never exceeds 40% of the total
surface class I molecules (Machy et al, 1987b). There are no changes in the
rate or extent of this internalization following cross linking of the mole-

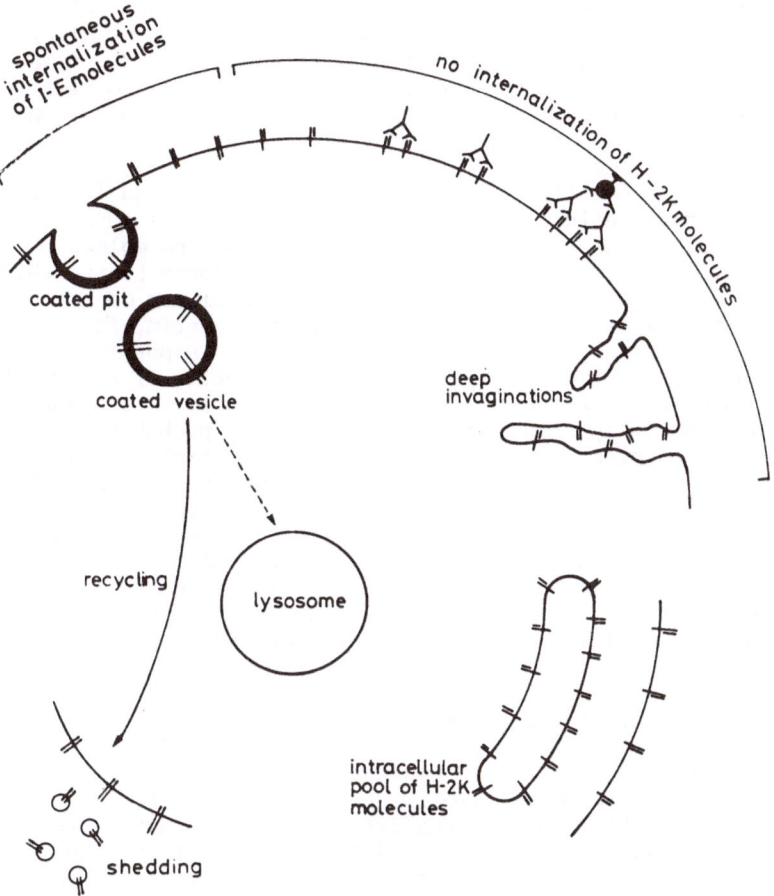

Fig. 3.   Behavior of MHC class I and II molecules in B lymphocytes.
$\restriction$ Class I H-2K molecules, $\curlyvee$ Anti-H-2K antibody,   ⊶ Cross
linker (protein A-bearing liposomes),   ‖ Class II I-E molecules.

cules with multivalent ligands. Furthermore, internalization occurs via
coated pits and coated vesicles and the internalized molecules are degraded
with a half life of about five hours (Machy et al, 1987b; Machy and Truneh,
1989). In general, lymphocytes have a very poor phagocytic capability as
already mentioned (Goldmacher et al, 1986). This is in agreement with our
observations that they cannot endocytose liposomes larger than 100-150 nm,
which corresponds exactly to the size of the coated pits by which we have
observed entry of the target molecules (Machy et al, 1987b; summarized in
Fig. 2).

Thus, the same molecule behaves differently when expressed by different
cell types. Nevertheless, targeting biologically active molecules to fibroblasts or T lymphocytes should not be a problem if MHC class I molecules are
the target cell surface determinants. In both cases, MTX-containing small
liposomes are delivered inside these cells and MTX inhibits specifically
cell proliferation. However, if we analyze another cell type, B lymphocytes, the results can be summarized as follows: no internalization of class
I molecules is observed and their half life on the cell surface is greater
than twenty hours (Machy and Truneh, 1989); no internalization is induced by
cross linking the target MHC class I molecules (Machy et al, 1987b); thus,
little or no internalization of the liposomes occurs (Leserman et al, 1981;

Machy et al, 1982a, b). If MTX-containing liposomes are used, only slight inhibition of cell proliferation will be observed (Leserman et al, 1981; Machy et al, 1982a, b). This pathway is summarized in Fig. 3. Based on these results, in all targeted systems used it is necessary to take into account the induced or intrinsic potential for internalization of the target molecule.

Do B cells endocytose less, in general, than T cells and fibroblasts? In the same vein, do T cells constitutively internalize all their surface molecules and do fibroblasts generally internalize surface molecules after cross linking? The answers are "No", and this corresponds to the third point we would like to emphasize. For example, fibroblasts do not constitutively internalize either class I molecules or fibronectin, for which they have receptors. Nevertheless, after cross linking it is possible to induce internalization of class I molecules but not fibronectin, if it is taken as the target (Truneh et al, 1983b). In contrast, fibroblasts are able to internalize constitutively transferrin receptors, and probably many other receptors, via coated pits (for review see Goldstein et al, 1985). For T lymphocytes, Thy-1 molecules are not internalized whether cross-linked or not (our unpublished results). Finally, for B lymphocytes there is no internalization of class I molecules as mentioned, but B lymphocytes do internalize their MHC class II molecules. This example has been further investigated because of our interest in the function of class II molecules in antigen presentation (Machy et al, 1989). We observed that liposomes directed against class II I-A and I-E molecules are internalized (Leserman et al, 1981; Machy et al, 1982a, b), but that internalization is dependent upon the site on the molecules being recognized by the anti-class II antibodies used (Machy et al, 1982b). This was not the case for the class I molecules in T lymphocytes (Machy et al, 1982b) or fibroblasts (Truneh et al, 1983a) where antibodies directed against different parts of the molecules were identical in terms of liposome endocytosis. For class II I-E molecules in B lymphocytes, the internalization occurs constitutively via coated pits/coated vesicles with a half life of thirty minutes and the molecules are recycled back to the cell surface and shed into the supernatant in vesicles of about 200 nm in diameter (Machy et al, 1989). This is also summarized in Fig. 3. What happens when liposomes are directed against regions of the molecule which are ineffective for internalization remains obscure. Possibly, cross linking of the part of the molecules against which liposomes are directed blocks the endocytic process because of the size of such generated complexes. This is in agreement with our unpublished observations that extensive cross linking of class II molecules slows down the rate of internalization. There is, however, no direct evidence that these determinant-specific complexes exist, and other explanations are not excluded. Thus, the behavior of the liposomes bound to the cell surface is controlled by the behavior of the target molecule and the way this target molecule behaves as a consequence of binding the targeted liposomes.

What happens to the liposomes and their contents once inside the cell? Do liposomes always release their encapsulated or lipid-associated materials in intact form, so that these materials mediate their biological effects? Our studies indicate that, once internalized, liposomes are delivered into acidic compartments such as endosomes, and end up in late endosomes or lysosomes. We have followed these pathways by evaluating the release of self-quenched carboxyfluorescein from liposomes into the cells, and by electron microscopy of liposomes containing horseradish peroxidase. MHC class I molecules were the target. Leakage inside the cell of carboxyfluorescein from liposomes increases the cells' fluorescence intensity, due to the dilution of the carboxyfluorescein in the cytoplasm (Truneh et al, 1983a, b). This release is probably the result of the neutralization of charges of the carboxyfluorescein in acidic compartments (endosomes and lysosomes) which enhances its leakage from intact liposomes (Barbet et al,

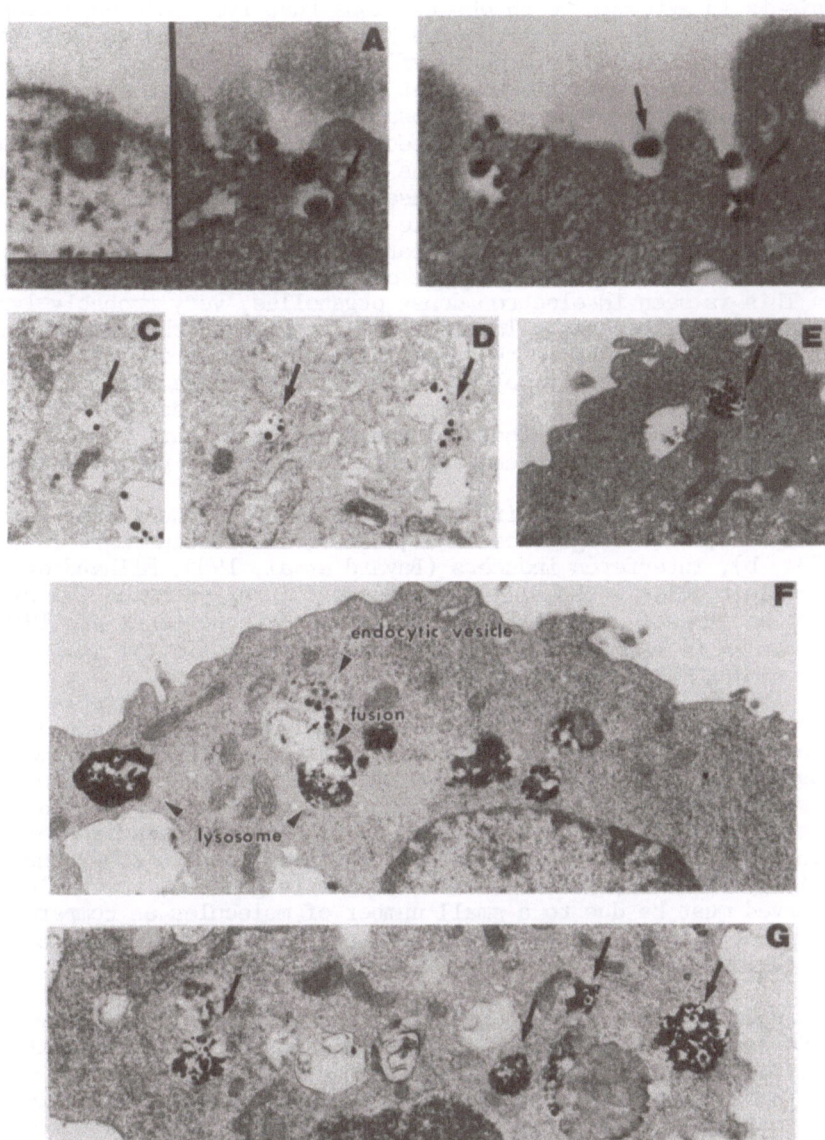

Fig. 4. Electron microscope studies of liposome entry into L cells via
class I molecules. L cells were incubated at 4°C with anti-H-2K
monoclonal antibodies and protein A-bearing liposomes containing
horseradish peroxidase (Machy et al, 1987a).
A and B, after 1 min incubation at 37°C, liposomes were found in
smooth endocytic vesicles, even though the cells had many coated
pits (insert in Fig. 4A). Magnification: 100,000x.
C, D and E, Time-dependent concentration of endocytosed intact
liposomes in endosomes. Fig. 4E corresponds to 5 min. Magnification 40,000x.
F, Time 30 min, liposomes are located in electron dense organelles
(lysosomes) where the punctate staining is replaced by a diffuse
staining. Mangification: 20,000x.
G, Time 2 h, intact liposomes are no longer seen. The cell-bound
materials have been internalized and are in lysosomal compartments.
Magnification: 20,000x.

1984). This is in agreement with observations that lysosomotropic compounds such as ammonium chloride or chloroquine that neutralize the acidity of cellular organelles inhibit both carboxyfluorescein leakage (Truneh et al, 1983b) and the effect of MTX (Leserman et al, 1981; Machy et al, 1982a). Looking at the endocytic pathway of liposome entry in fibroblasts (Fig. 4), we can see that liposomes are internalized in intact form, demonstrating definitely that they do not fuse with the plasma membrane. In this example, liposomes contained horseradish peroxidase. After internalization at 37°C, intact liposomes are present in endocytic vesicles and over time they are concentrated, still in intact form, in larger endosomes. With longer incubation the punctate staining is no longer observed but is replaced by diffuse staining. This is seen in electron dense organelles, very probably lysosomes, containing cell debris. This diffuse staining is the consequence of release of liposome contents in these cellular compartments. At no time did we observe labelling of the cytoplasm. The punctate (intact liposomes) or diffuse (degraded liposomes) were detected only inside endosomes or lysosomes respectively, suggesting that for the peroxidase there was little or no diffusion from the endocytic compartments to the other parts of the cell.

So far, we have delivered different soluble drugs (Machy and Leserman, 1984), lipid-associated drugs (Noe et al, 1988), fluorescent markers (Truneh et al, 1983a, b), interferon inducers (Bayard et al, 1985; Milhaud et al, 1989) and nucleic acids such as marker gene-containing plasmids (Machy et al, 1988) or anti-sense oligodeoxyribonucleotides (Leonetti et al, 1989) into target cells with retention of their biological activity. Based on the finding concerning the behavior of peroxidase we may ask how and in what quantity a given molecule reaches its site of action in the cytoplasm or in the nucleus. What is the proportion of these molecules concentrated in endosomes and lysosomes that can escape in intact form from these compartments? Based on our results with the horseradish peroxidase-containing liposomes we may assume that this proportion may be very low. It is, nevertheless, possible that what we observed with the horseradish peroxidase is only true for that molecule and cannot be generalized to other encapsulated materials. However, if true for molecules in general, the biological effect observed must be due to a small number of molecules as compared to the number of molecules in liposomes at the cell surface, since we can expect eventual degradation of liposomes in lysosomes.

What happens to liposome-encapsulated molecules once in lysosomes undoubtedly depends on the nature of the molecule in question. This is not necessarily "fatal" for carboxyfluorescein and MTX, since their low molecular weight and lipophilicity at acid pH presumably permit their release into the cytoplasm. For a protein however, passage in the lysosome could be a problem. It is thought, for example, that the fraction (about 5%) of bound immunotoxins which escape the lysosomal system is responsible for the cytotoxicity of the reagent, and the rest is degraded (for review see Olsnes et al, 1989). Apparently, degradation in lysosomes depends on the molecules in question since for horseradish peroxidase there is no degradation for at least seven hours, as determined by the electron density observed in electron microscopy, which depends on the enzymatic activity of the molecule. On the other hand, enzymatic cleavage may be necessary for some molecules to be active. This is the case for the lipid derivatized MTX: MTX-γ-dimyristoyl-phosphatidylethanolamine, for which cleavage by endosomal or lysosomal phospholipases may permit release of MTX from the lipids and allow the drug to reach the cytoplasm (Noé et al, 1988). In conclusion, the delivery of the liposome contents may be highly dependent on the nature of these contents, their susceptibility to acidic pH in endosomes and to enzymatic cleavage in lysosomes.

To conclude this chapter, in all targeted systems developed for therapeutic or cell biological purposes it is important to consider the target molecule, the cell that expresses it, the nature of the encapsulated molecule and the site it must reach inside the cell. We believe that our understanding of the physiology of the endocytic process and the mechanisms by which the encapsulated molecules are released from the liposomes and the intracellular compartments will enhance considerably the desired biological effects. In addition, improvement in liposome technology should be of major interest to permit delivery in intact form of the encapsulated material inside the cell without delivery to lysosomal compartments, particularly for molecules which are degraded by lysosomal enzymes. The chemistry of encapsulated molecules could be of great importance in avoiding degradation.

Acknowledgements

We are grateful to Dr. Alemseged Truneh for valuable exchanges, and to Drs. Sylvia Hoffstein and Hubert Reggio as well as to Doris Gennaro and Jean Pierre Bizozzero for their expertise in electron microscopy.

REFERENCES

Barbet, J., Machy, P. and Leserman, L.D., 1981, Monoclonal antibody covalently coupled to liposomes: specific targeting to cells, J.Supramol.Struct. and Cell Biochem., 16:243.

Barbet, J., Machy, P., Truneh, A. and Leserman, L.D., 1984, Weak acid-induced release of liposome-encapsulated carboxyfluorescein, Biochim.Biophys.Acta, 772:347.

Bayard, B., Leserman, L.D., Bisbal, C. and Lebleu, B., 1985, Antiviral activity in L1210 cells of antibody-targeted liposomes containing (2'-5')oligo(adenylate) analogues, Eur.J.Biochem., 151:319.

Goldmacher, V.S., Tinnel, N.L. and Nelson, B.C., 1986, Evidence that pinocytosis in lymphoid cells has a low capacity, J.Cell Biol., 102:1312.

Goldstein, J.L., Brown, M.S., Anderson, R.G.W., Russell, D.W. and Schneider, W.J., 1985, Receptor-mediated endocytosis: concepts emerging from the L.D.L. receptor system, Ann.Rev.Cell Biol., 1:1.

Huet, C., Ash, J.F. and Singer, S.J., 1980, The antibody-induced clustering and endocytosis of HLA antigens on cultured human fibroblasts, Cell, 21:429.

Leonetti, J-P., Machy, P., Degols, G., Lebleu, B. and Leserman, L.D., 1989, Antibody-targeted liposomes containing oligodeoxyribonucleotide sequences complementary to viral RNA selectively inhibit viral replication, Submitted for publication.

Leserman, L. and Machy, P., 1987, Ligand targeting of liposomes, in "Liposomes: from Biophysics to Therapeutics", M.J. Ostro, ed., Marcel Dekker, Inc., New York.

Leserman, L.D., Barbet, J., Kourilsky, F. and Weinstein, J.N., 1980, Targeting to cells of fluorescent liposomes covalently coupled with monoclonal antibody or protein A, Nature, 288:602.

Leserman, L.D., Machy, P. and Barbet, J., 1981, Cell specific drug transfer from liposome bearing monoclonal antibodies, Nature, 293:226.

Machy, P. and Leserman, L.D., 1983, Small liposomes are better than large liposomes for specific drug delivery in vitro, Biochim.Biophys.Acta, 730:313.

Machy, P. and Leserman, L.D., 1984, Elimination or rescue of cells in culture by specifically targeted liposomes containing methotrexate or formyl tetrahydrofolate, EMBO J., 3:1971.

Machy, P. and Leserman, L.D., 1987, "Applications of Liposomes in Cell Biology and Pharmacology", Editions INSERM, Paris.

Machy, P. and Truneh, A., 1989, Differential half life of major histocompatibility complex encoded class I molecules by T and B lymphoblasts, Mol.Immunol., 26:687.

Machy, P., Barbet, J. and Leserman, L.D., 1982a, Differential endocytosis of T and B lymphocyte surface molecules evaluated with antibody fluorescent liposomes containing methotrexate, Proc.Natl.Acad.Sci. USA, 79:4148.

Machy, P., Pierres, M., Barbet, J. and Leserman, L.D., 1982b, Drug transfer into lymphoblasts mediated by liposomes bound to distinct sites on H-2 encoded I-A, I-E and K molecules, J.Immunol., 129:2098.

Machy, P., Truneh, A., Gennaro, D. and Hoffstein, S., 1987a, Endocytosis and de novo expression of major histocompatibility complex encoded class I molecules: kinetic and ultrastructural studies, Eur.J.Cell Biol., 45:126.

Machy, P., Truneh, A., Gennaro, D. and Hoffstein, S., 1987b, Major histocompatibility complex class I molecules internalized via coated pits in T lymphocytes, Nature, 328:724.

Machy, P., Lewis, F., McMillan, L. and Jonak, Z., 1988, Gene transfer from targeted liposomes to specific lymphoid cells by electroporation, Proc.Natl.Acad.Sci.USA, 85:8027.

Machy, P., Bizozzero, J-P., Reggio, H. and Leserman, L., 1989, Endocytosis, recycling and shedding of major histocompatibility complex-encoded class II molecules by murine B lymphocytes. Submitted for publication.

Milhaud, P.G., Machy, P., Lebleu, B. and Leserman, L., 1990, Antibody targeted liposomes containing poly (rI).poly (rC) exert a specific antiviral and toxic effect on cells primed with interferons α/β or γ, Biochim.Biophys.Acta, in press.

Noé, C., Hernandez-Borrell, J., Kinsky, S.C., Matsuura, E. and Leserman, L., 1988, Inhibition of cell proliferation with antibody-targeted liposomes containing methotrexate-γ-dimyristoylphosphatidylethanolamine, Biochim.Biophys.Acta, 946:253.

Olsnes, S., Sandvig, K., Petersen, O.W. and van Deurs, B., 1989, Immunotoxins - entry into cells and mechanisms of action, Immunol.Today, 10:291.

Ostro, M.J., ed., 1987, "Liposomes: from Biophysics to Therapeutics", Marcel Dekker, Inc., New York.

Truneh, A. and Machy, P., 1987, Detection of very low receptor numbers on cells by flow cytometry using a sensitive staining method, Cytometry, 8:562.

Truneh, A., Machy, P., Barbet, J., Mishal, Z., Lemonnier, F.A. and Leserman, L.D., 1983a, Endocytosis of HLA and H-2 molecules on transformed murine cells measured by fluorescence dequenching of liposome-encapsulated carboxyfluorescein, EMBO J., 2:2285.

Truneh, A., Mishal, Z., Barbet, J., Machy, P. and Leserman, L.D., 1983b, Endocytosis of liposomes bound to cell surface proteins measured by flow cytometry, Biochem.J., 214:189.

Truneh, A., Mishal, Z. and Leserman, L.D., 1984, A calmodulin antagonist increases the rate of endocytosis of liposomes bound to MHC molecules via monoclonal antibodies, Exp.Cell Res., 155:53.

Truneh, A., Machy, P. and Horan, P.K., 1987, Antibody-bearing liposomes as multicolor immunofluorescence markers for flow cytometry and imaging, J.Immunol.Methods, 100:59.

TISSUE SPECIFIC SERUM OPSONINS AND PHAGOCYTOSIS OF LIPOSOMES

Harish M. Patel and S. Moein Moghimi

Department of Biochemistry, Charing Cross and Westminster
Medical School, Fulham Palace Road, London W6 8RF, UK

INTRODUCTION

Blood clearance of intravenously injected colloidal particles such as drug carriers, liposomes, nanoparticles, microspheres etc. occurs mainly in the liver, spleen and bone-marrow by phagocytes lining blood sinuses and of these sites hepatic clearance by Kupffer cells predominates. The natural homing of drug-carriers such as liposomes to elements of the reticulo-endothelial system (RES) provides the advantage of specific targeting of drugs and immunomodulators to macrophages (Alving, 1982; Poste et al, 1979). However, it limits the prospect of delivery of drugs to cells other than phagocytic cells of liver, spleen and bone-marrow and thus the RES proves to be a major obstacle in the targeting of liposomes to other cell types and tissues. Attempts to divert liposomes away from the RES have been made by 'blockading' the phagocytic uptake mechanisms by pretreating animals with substances such as carbon particles or dextran sulphate (Souhami et al, 1981; Freise et al, 1981) or by saturating macrophages by predosing animals with empty liposomes followed by the 'test' liposomes (Abra et al, 1980; Dave and Patel, 1986). These manipulations have limited benefits in pro-longing circulating half-life and alteration of tissue distribution, partic-ularly among the RES organs. The half-life of liposomes in the circulation can also be prolonged by altering the lipid composition of liposomes (Gregoriadis and Senior, 1980; Senior and Gregoriadis, 1982; Gabizon and Papahadjopoulos, 1988; Allen et al, 1989). For example, liposomes contain-ing cholesterol and prepared from sphingomyelin or saturated phospholipids such as dimyristoyl phosphatidylcholine, dipalmitoyl phosphatidylcholine or distearoyl phosphatidylcholine have an extremely long half-life in the circulation (Gregoriadis and Senior, 1980; Senior and Gregoriadis, 1982). Similarly, inclusion of certain gangliosides or phosphatidylinositol which gives the liposomal surface a negative charge and increased hydrophilicity prolong the circulating half-life of liposomes with a concomitant decrease in liver and spleen uptake (Gabizon and Papahadjopoulos, 1988; Allen et al, 1989).

However, it is not understood why these lipids cause a change in the clearance rate of liposomes. The clearance of liposomes depends partly on their stability but mainly on phagocytic activity of mononuclear phagocytes of the RES (Senior and Gregoriadis, 1982; Patel et al, 1983). This argument is further supported by the fact that colloidal particles such as carbon, metal colloid, gold and polystyrene particles are stable and yet rapidly

Targeting of Drugs, Edited by G. Gregoriadis *et al.*
Plenum Press, New York, 1990

cleared from the circulation. The rate of clearance of liposomes and other materials by the RES depends on their surface properties such as charge, size, fluidity and degree of hydrophobicity or hydrophilicity and these properties can be altered by selecting the lipid compositions of liposomes. These properties play an important role in attracting various serum components which may act as opsonins or dysopsonins and thus influence liposome phagocytosis. So far, however, studies carried out <u>in vitro</u> have not established the opsonic function of serum in the phagocytosis of liposomes (Juliano, 1982; Dijkstra et al, 1984; Ellens et al, 1981, 1982).

Recently, we have demonstrated that intravenously injected cholesterol-free and cholesterol-containing liposomes are handled differently by liver and spleen (Patel et al, 1983; Dave and Patel, 1986). Liver takes up cholesterol-free more than cholesterol-containing liposomes, whereas spleen prefers cholesterol-containing more than cholesterol-free liposomes. Hence, to understand the mechanism of differential uptake of these liposomes by liver and spleen macrophages, we examined the role of serum on phagocytosis of liposomes by phagocytic cells from liver, spleen, bone marrow and lung.

METHODOLOGY

Negatively charged multilamellar liposomes with entrapped C-14-inulin or I-125-poly vinylpyrrolidone (PVP) were prepared from a mixture of phosphatidylcholine/cholesterol/dicetyl phosphate in a molar ratio of 7:0:1 for cholesterol free, 7:2:1 for cholesterol poor, and 7:7:1 for cholesterol rich liposomes, respectively. Hepatic non-parenchymal cells containing predominantly endothelial and Kupffer cells and splenic white cells were prepared from male CFY rats of body weight 250±25 g. Cells were incubated at 37°C for one hour in the absence or presence of 25% rat serum. Liver cells (10^7) were incubated in Ca^{2+}/Mg^{2+} free Hank's balanced salt solution (Gibco) and spleen cells (8^7-10^7) were incubated in 10mM phosphate-saline buffer, pH 7.4. The opsonic activity in serum is expressed as uptake of liposomes by cells monitored at the end of the incubation by measurement of the radioactivity of liposomal I-125-PVP or C-14-inulin associated with cells. Results are expressed as percentage uptake of the total radioactivity in the incubation mixture (mean of three samples ± SD).

SERUM OPSONIC ACTIVITY

Results in Fig. 1 show that in the absence of serum, liver and spleen cells take up cholesterol-free more than cholesterol-containing liposomes. Among cholesterol-containing liposomes, cholesterol-poor are taken up much more by liver and spleen cells than cholesterol-rich liposomes. When serum is included in the incubation mixture, these results are altered indicating that serum plays a role in the uptake of liposomes by liver and spleen phagocytes. With both liver and spleen cells, serum suppresses the uptake of cholesterol-free liposomes. This could be due to their rapid degradation in serum (Kirby et al, 1980; Patel et al, 1983), which may result in a smaller number of intact liposomes available to cells for phagocytosis. However, serum has a variable effect on the uptake of liposomes depending on their cholesterol content and source of macrophages. It enhances the uptake of cholesterol-poor but suppresses that of cholesterol-rich liposomes in liver cells (Fig. 1) but it stimulates the uptake of these liposomes in spleen cells with a greater opsonic effect on the uptake of cholesterol-rich than cholesterol-poor liposomes. These results are comparable with those of <u>in vivo</u> experiments reported earlier (Patel et al, 1983; Dave and Patel, 1986) and suggest that serum contains opsonins specific for liver and spleen phagocytic cells and these opsonins have different affinity for cholesterol-poor and cholesterol-rich liposomes.

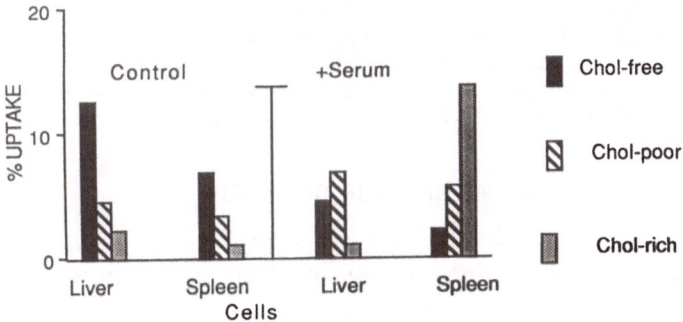

Fig. 1.   Effect of serum on the uptake of negatively charged MLV egg PC
liposomes in liver and spleen cells.

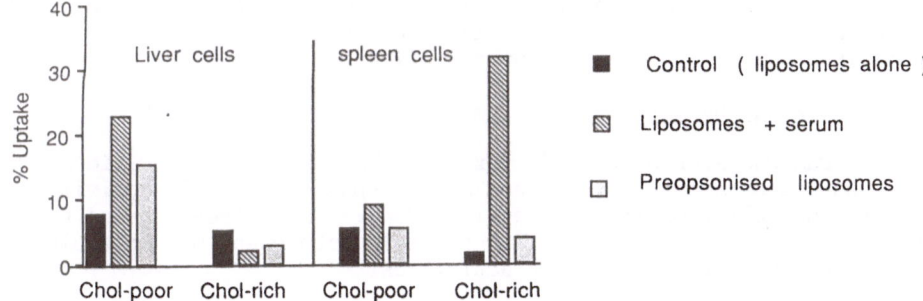

Fig. 2.   Uptake of pre-opsonized negatively charged MLV liposomes in liver
and spleen cells.

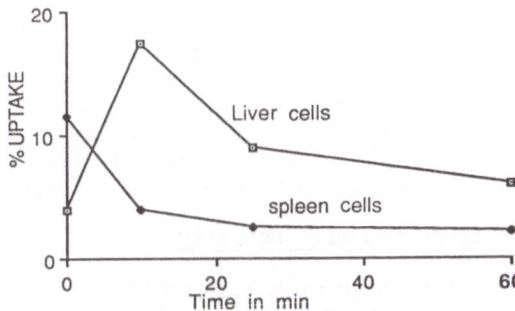

Fig. 3.   Effect of heat treatment of serum at 55°C on its opsonic activity.

Next, cholesterol-poor and cholesterol-rich liposomes were pre-opsonised by incubating in 50% serum at 37°C for 15 min (Moghimi and Patel, 1988)
and their uptake by liver and spleen cells was studied.   Results in Fig. 2
show that opsonisation of cholesterol-poor enhanced their uptake, but
suppressed that of cholesterol-rich liposomes in liver cells.   Washing of
cholesterol-poor pre-opsonised liposomes prior to incubation with liver
cells caused a loss of the opsonic activity (Moghimi and Patel, 1988).   On
the other hand, the opsonised cholesterol-poor and cholesterol-rich lipo-

somes have no opsonic effect on their uptake by spleen cells. These results probably suggest that liver specific opsonin mediate its opsonic action by being adsorbed loosely on the surface of cholesterol-poor liposomes whereas spleen specific opsonin does not appear to mediate its function by binding onto the surface of liposomes (Moghimi and Patel, 1988).

PROPERTIES OF LIVER AND SPLEEN SPECIFIC OPSONINS

To establish that liver and spleen specific opsonins are two separate molecules, some of the basic properties of these opsonins were studied.

Heat Stability

Results in Fig. 3 show that liver specific activity was increased initially and reached a maximum level when serum was preheated at 55°C for 10 min. Thereafter, further heating of serum caused a gradual decrease in the activity and even after prolonged heating for 60 min, the activity was still elevated as compared to that of unheated serum. The spleen specific opsonin, on the other hand, was gradually lost on preheating serum at 55°C. From these results we concluded that liver specific opsonin is heat stable whereas spleen specific opsonin is heat labile.

Effect of Freezing and Thawing of Serum

When serum was stored at -20°C for 6 months, its liver specific activity was enhanced by about 25%, whereas spleen specific opsonic activity was reduced by about 20% (Moghimi and Patel, 1989a). Thus, freezing and thawing probably brings about conformational changes in both opsonins' molecules which have reverse effects on their activities.

Effect of Dialysis of Serum

When serum was dialysed, it produced opposite effects on the activity of liver and spleen specific opsonins. Liver-specific activity was enhanced by about two fold, whereas spleen specific activity was reduced by about 50% of control serum (Fig. 4). The dialysate which was collected at the end of dialysis and concentrated to the original volume of serum used for dialysis had no effect on the uptake of liposomes in liver and spleen cells when included in the incubation. But when this dialysate was added back to dialysed serum, the elevated liver specific opsonin activity of the dialysed serum was suppressed almost to the level of that of normal serum, whereas it failed to restore spleen specific activity to the level of normal serum. Thus these results indicated that serum contains a dialysable factor(s) which regulated the activities of both opsonins with opposite effect.

Dialysable Factor and Liver Specific Activity

When fresh serum was treated with 1.25 mM EGTA, its liver specific activity was enhanced two to three fold (Fig. 5) and was comparable to the activity found in dialysed serum (Moghimi and Patel,1989c). Dialysis of serum removes small molecular weight substances including ions and EGTA is known to chelate divalent cations. Thus these results suggest that some divalent cations may be involved in the regulatory mechanism of the serum's liver specific opsonic activity. Hence the effect of various divalent cations such as CO^{2+}, Mn^{2+}, Mg^{2+}, Ca^{2+} on the uptake of liposomes by liver cells was studied (Moghimi and Patel, 1989c). These cations at concentration of 1.5 mM suppressed the opsonic activity of dialysed serum since the uptake of liposomes by liver cells was greatly reduced. However, considering the physiological concentration and the amount of these cations used in our experiments, it became clear that calcium may be the important

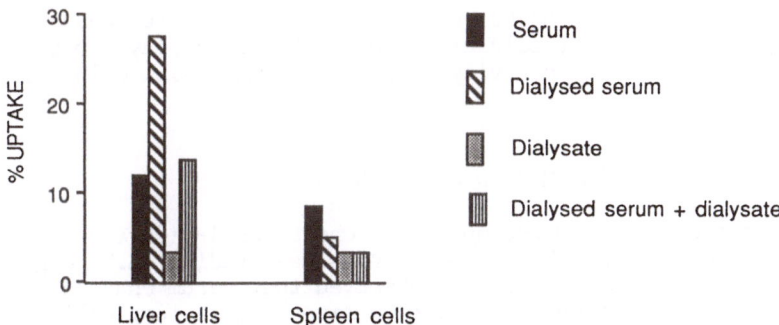

Fig. 4. Opsonic activity in dialysed rat serum.

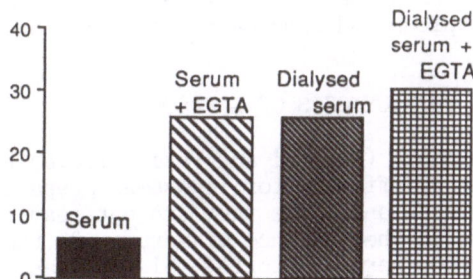

Fig. 5. Effect of EGTA on liver specific opsonic activity in control and dialysed serum.

regulatory factor for liver specific activity. The calcium level in our rat serum, measured by atomic absorption, indicated that dialysis removes about 1.5 mM calcium from a total of 2.6 mM (average) in normal serum. Results in Fig. 6 show that the enhanced activity observed in dialysed serum was gradually reduced when calcium chloride was added in the incubation mixture. 1.8 mM calcium chloride brings the calcium level of the dialysed serum back to that of normal serum. The effect of calcium chloride on the uptake of

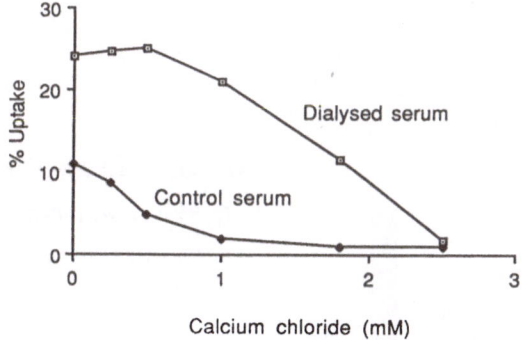

Fig. 6. Effect of calcium chloride on liver opsonic activity in dialysed and control serum. Negatively charged MLV cholesterol-poor liposomes were used for the uptake study.

liposomes by spleen cells in the presence of dialysed serum was also studied. It was found that calcium (calcium chloride) plays no role in regulating spleen specific activity. Similarly, magnesium chloride has no effect on it.

OPSONINS SPECIFIC FOR BONE MARROW AND LUNG MACROPHAGES

The effect of serum on the uptake of cholesterol-rich and cholesterol-poor liposomes by lung and bone marrow cells was studied (S.M. Moghimi and H.M. Patel, in preparation). It was observed that serum enhances uptake of cholesterol-poor but not cholesterol-rich liposomes in lung cells, whereas it enhances phagocytosis of both cholesterol-poor and cholesterol-rich liposomes by bone marrow. The lung-specific opsonin is heat stable and its activity is enhanced when serum is dialysed, whereas bone marrow-specific opsonin is heat labile since its activity is reduced on heating serum at 55°C. Its activity is also partially lost when serum is dialysed. Thus our results indicate that the properties of liver and lung specific opsonins are similar and those of spleen and bone marrow are similar.

SERUM DYSOPSONINS AND PHAGOCYTOSIS OF LIPOSOMES

Recently we have shown (Moghimi and Patel, 1989b) that liver and spleen specific opsonins have no affinity for liposomes prepared from sphingomyelin or saturated phospholipids and hence they are not readily phagocytosed by liver and spleen cells in the presence of serum. On the contrary, these liposomes attract unknown serum factor(s) which act as dysopsonins specifically for liver cells and thus inhibit their uptake by these cells but do not affect uptake by spleen cells. This suggests that serum may also contain dysopsonins which may show tissue specificity. This is further supported by the finding that inclusion of cholesterol (46.6%) in these liposomes improves their uptake by spleen cells (Fig. 7) but not by liver cells.

CONCLUSION

Our results obtained with freshly prepared cells are very similar to those obtained with in vivo studies (Patel et al, 1983; Dave and Patel, 1986) and provide evidence for the first time that serum contains tissue specific opsonins which enhance phagocytosis of liposomes by liver, spleen, bone marrow and lung macrophages. The opsonic effect of serum depends on lipid composition and surface properties such as fluidity, charge, hydro-

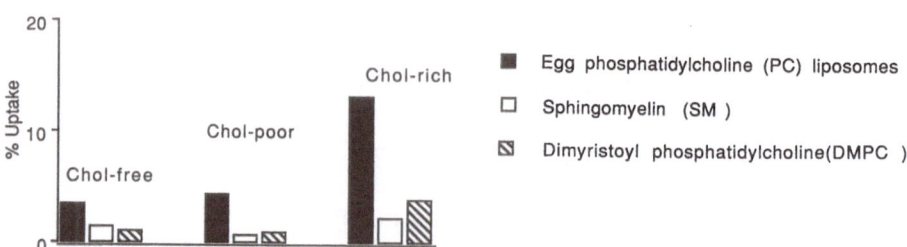

Fig. 7. Effect of serum on uptake of cholesterol containing PC, SM and DMPC negatively charged MLV liposome in spleen cells.

phobicity/hydrophilicity of lipid bilayers of liposomes which play an important role in attracting these opsonins or dysopsonins. For example, cholesterol-rich liposomes have greater affinity for spleen specific opsonins whereas they have poor affinity for liver specific opsonin. Similarly serum has no opsonic effect on liposomes prepared from sphingomyelin and saturated phospholipids and this suggests that these liposomes have no affinity for serum opsonins. The word 'affinity' has a wider meaning here, since there can be more than one possible reason why serum does not exert its effect on phagocytosis of these liposomes by liver and spleen macrophages. The first possibility is that tissue specific opsonins may not interact with liposomes whose surface properties are not favourable for binding of opsonins. The second possibility is that opsonins may interact with the liposome surface in such a way that their 'receptor binding site' may not be exposed correctly to interact with macrophages or that other serum proteins and components which may interact with liposomes may overshadow this site. Further experiments should be designed to understand the mechanisms of binding tissue specific opsonins and dysopsonins onto liposomes with different properties.

So far we have not identified any of these tissue specific opsonins or dysopsonins but we know that liver-specific opsonin is not an immunoglobulin since during our attempts to purify this opsonin, its activity did not precipitate with immunoglobulins in ammonium sulphate fraction. It cannot be fibronectin because liver opsonin is heat stable and its activity is enhanced on freezing and thawing, whereas fibronectin is heat labile and loses it activity on cold storage and freezing and thawing of serum. Furthermore fibronectin depleted serum was found to retain its opsonin activity (Moghimi and Patel, 1989a). We have speculated earlier (Moghimi and Patel, 1988; Moghimi and Patel, 1989a) that serum contains more than one spleen specific opsonin which may exert its effect together or individually depending on liposome preparations and physiological or experimental conditions. The properties of splenic opsonin described as discussed here and elsewhere (Moghimi and Patel, 1988; 1989a,b,c) suggest to us to consider the possibility of fibronectin and complenent factor C3b as joint components of spleen specific opsonin.

Lastly, we believe that purification and characterisation of both tissue specific opsonins and dysopsonins will provide us with an opportunity to target drug carriers selectively to a particular organ of the reticuloendothelial system and also help us to divert carriers away from the RES in order to prolong their circulating half-life in circulation. Moreoever, understanding the regulatory mechanism of tissue opsonins may help us to evaluate their role in infection and other pathological conditions.

REFERENCES

Abra, R.M., Bosworth, M.E. and Hunt, A.C., 1980, Liposome disposition in vivo: effect of pre-dosing with liposomes, Res.Commun.Chem.Pathol. Pharmacol., 29:349.

Allen, T.M., Hansen, C. and Rutledge, J., 1989, Liposomes with prolonged circulation times: factors affecting uptake by reticuloendothelial and other tissues, Biochim.Biophys.Acta, 981:27.

Alving, C.R., 1982, Therapeutic potential of liposomes as carriers in leishmaniasis, malaria, and vaccines, in "Targeting of Drugs", G. Gregoriadis, J. Senior and A. Trouet, eds., Plenum Press, N.Y.

Dave, J. and Patel, H.M., 1986, Differentiation in hepatic and splenic phagocytic activity during reticuloendothelial blocked with cholesterol-free and cholesterol-rich liposomes, Biochim.Biophys. Acta, 888:184.

Dijkstra, J., van Galen, W.J.M., Hulstaert, C.E., Kalicharen, D., Roerdick,

F.H. and Scherphof, G.L., 1984, Interaction of liposomes with Kupffer cells in vitro, Exp.Cell Res., 150:161.

Ellens, H., Morselt, H. and Scherphof, G., 1981, In vivo fate of large uni-lamellar sphingomyelin cholesterol liposomes after intraperitoneal and intravenous injection into rats, Biochim.Biophys.Acta, 674:10.

Ellens, H., Mayhew, E. and Rustum, Y.M., 1982, Reversible depression of the reticuloendothelial system by liposomes, Biochim.Biophys.Acta, 714:479.

Freise, J., Müller, W.H. and Magerstedt, P., 1981, Uptake of liposomes and sheep red blood cells by the liver and spleen of rats with normal decreased function of the reticuloendothelial system, Res.Expt.Med. (Berlin), 178:263.

Gabizon, A. and Papahadjopoulos, D., 1988, Liposome formulations with pro-longed circulation time in blood and enhanced uptake by tumors, Proc.Natl.Acad.Sci.USA, 85:6949.

Gregoriadis, G. and Senior, J., 1980, The phospholipid component of small unilamellar liposomes controls the rate of clearance of entrapped solutes from the circulation, FEBS Lett., 119:43.

Juliano, R.L., 1982, Liposomes and the reticuloendothelial system: interactions of liposomes with macrophages and behaviour of liposomes in vivo, in "Targeting of Drugs", G. Gregoriadis, J. Senior and A. Trouet, eds., Plenum Press, New York.

Kirby, C., Clarke, J. and Gregoriadis, G., 1980, Effect of the cholesterol content of small unilamellar liposomes on their stability in vivo and in vitro, Biochem.J., 186:591.

Moghimi, S.M. and Patel, H.M., 1988, Tissue specific opsonins for phagocytic cells and their different affinity for cholesterol-rich liposomes, FEBS Lett., 233:143.

Moghimi, S.M. and Patel, H.M., 1989a, Differential properties of organ specific serum opsonins for liver and spleen macrophages, Biochim.Biophys.Acta, 984:379.

Moghimi, S.M. and Patel, H.M., 1989b, Serum opsonins and phagocytosis of saturated and unsaturated phospholipid liposomes, Biochim.Biophys. Acta, 984:384.

Moghimi, S.M. and Patel, H.M., 1989c, Ca^{2+} as regulatory factor in serum liver specific opsonic activity, Biochim.Biophys.Acta, submitted.

Patel, H.M., Tuzel, N.S. and Ryman, B.E., 1983, Inhibitory effect of cholesterol on the uptake of liposomes by liver and spleen, Biochim.Biophys.Acta, 761:142.

Poste, G., Kirsh, R., Fogler, W.E. and Fidler, I.J., 1979, Activation of tumoricidal properties in mouse macrophages by lymphokines encap-sulated in liposomes, Cancer Res., 39:881.

Senior, J. and Gregoriadis, G., 1982, Stability of small unilamellar lipo-somes in serum and clearance from circulation: the effect of phospho-lipid and cholesterol components, Life Sci., 30:2123.

Souhami, R.L., Patel, H.M. and Ryman, B.E., 1981, The effect of reticulo-endothelial blockade on the blood clearance and tissue distribution of liposomes, Biochim.Biophys.Acta, 674:354.

LIPOSOME TARGETING TO TUMOR CELLS IN VIVO

D. Papahadjopoulos[a] and A. Gabizon[a,b,c]

[a]Cancer Research Institute and Department of Pharmacology
University of California, San Francisco, California 94143
[b]Liposome Technology, Inc., Menlo Park, California 94025
[c]Present address: Hadassah Medical Center, Jerusalem 91120
Israel

INTRODUCTION

Considerable effort has been mounted both in our own and other laboratories in order to enhance the ability of liposomes to deliver drugs to specific cells. This effort resulted in the development of methods for conjugation of antibodies on the liposome surface (Heath et al, 1980; Leserman et al, 1980; Huang et al, 1980), temperature-sensitive liposomes (Papahadjopoulos et al, 1973; Yatvin et al, 1978) which can be used in conjunction with hyperthermia (Weinstein et al, 1979) and pH-sensitive liposomes, either for localization of drugs in low pH areas (Yatvin et al, 1980) or for release of the liposome-encapsulated drugs following endocytosis (Dusgunes et al, 1985; Ellens et al, 1984; Connor et al, 1984).

However, such systems could not be developed further into practical applications because of the rapid uptake of liposomes by the cells of the Reticuloendothelial System (RES) following their injection into the blood stream (Gregoriadis and Senior, 1986). The limited permeability of the endothelial barrier in most tissues to particles of the size usually encountered in liposome preparations would also tend to limit the possibility for targeting to cells outside the blood vasculature (Poste, 1983). Recent work, however, has shown that specific changes in the liposome composition can drastically reduce their uptake by the RES of liver and spleen (Allen and Chonn, 1987; Papahadjopoulos and Gabizon, 1987; Gabizon and Papahadjopoulos, 1988). Such liposomes circulated in the blood for much longer periods of time, compared to the usual compositions presently tested in clinical trials. In addition, intact liposomes have been observed in relatively high amounts in many tissues. Most importantly, they tend to accumulate in areas of implanted tumors at concentrations higher than any other organ or tissue of the body outside the RES (Papahadjopoulos and Gabizon, 1987). Furthermore, when liposomes are conjugated to monoclonal antibodies recognizing antigens on the tumor cell surface, there is a further increase in liposome uptake by the tumor, as opposed to other parts of the body (Gabizon and Papahadjopoulos, 1988). Such targeting effect to diseased parts of the body could revolutionize drug delivery for such life threatening diseases as cancer. However, much more work is needed before the results reported here can be transferred into medically relevant applications to humans.

Targeting of Drugs, Edited by G. Gregoriadis *et al.*
Plenum Press, New York, 1990

Table 1. Tissue Distribution of Liposomes in Normal Mice[a]

| Liposome Composition[b] (molar ratio) | Time after injection (h) | % of Recovered Dose per Compartment | | | |
		Blood	Liver + Spleen	Carcass + Skin	Other Organs
PG-PC-C (1-9-5)[c]	4	0.4	82.0	13.7	3.9
PG-PC-C (1-9-5)	4	5.9	70.6	19.6	3.9
GM_1-PC-C (1-9-5)	4	34.6	32.9	22.5	10.1
GM_1-PC-C (1-9-5)	24	8.7	48.8	31.4	11.1
DPPG-DSPC-C (1-10-5)	24	10.9	52.2	25.5	11.4
GM_1-DSPC-C (1-10-5)	24	24.2	25.5	35.7	14.6

[a]Dose: 1 μmole of phospholipid injected i.v. into Swiss Webster female mice. Results are the average of 3 mice.
[b]Except for preparation discussed under (c), all these preparations were extruded through a 0.05 or 0.1 μm polycarbonate membrane and had a particle size distribution of approximately 0.07 μm-0.12 μm.
[c]This preparation was composed of unextruded MLV with a particle size distribution from 5 to 0.5 microns in diameter.

TISSUE DISTRIBUTION OF LIPOSOMES IN NORMAL MICE: EFFECT OF LIPID COMPOSITION

The effect of lipid composition on tissue distribution in vivo was studied by screening a variety of liposome formulations for their localization in various tissues in mice after i.v. injection. All liposomes were made by thin lipid film hydration, followed by extrusion through polycarbonate membranes with pores of 0.08 or 0.05 μm-diameter as previously described (Gabizon and Papahadjopoulos, 1988). The vesicles were radio-labeled with gallium-67 deferoxamine according to a method previously reported (Gabizon et al, 1988). We have arbitrarily divided the body into four anatomic compartments (blood, liver-spleen, carcass-skin, other organs) and calculated for each of them the percentage of recovered dose. Blood represents the pool of circulating liposomes. Liver plus spleen is taken as an approximation of the uptake by the RES. Carcass plus skin includes all bones, muscles and skin. Other organs include gut and appendages, kidneys, heart and lungs.

The various factors that play a role in prolonging circulation time and improving biodistribution of liposomes are presented in Table 1. The percentage of recovered dose in the various body compartments was determined at 4 h and 24 h after injection of six representative formulations of liposomes. The effect of liposome configuration and size can be clearly recognized when comparing the unextruded MLV preparation of PG-PC-C (size distribution range 0.5-5.0 μm) to the extruded liposomes (mean size range 70-120 nm) of the same composition at 4 h after injection. A more than 10-fold increase in the amount remaining in blood is obtained. However, in both cases the majority of the label is recovered in liver-spleen. When the negatively-charged PG is replaced at the same molar ratio by a negatively-charged glycolipid, the ganglioside GM1, there is a drastic reduction in the amount accumulated in the liver plus spleen compartment and, concomitantly, the circulating liposome pool is significantly larger (about 6-fold) than with PG-containing liposomes.

Table 1 also shows the distribution of the GM1-PC-C liposomes 24 h after injection and compares them to two more preparations where PC (derived from egg yolk and forming a fluid bilayer at body temperature) is replaced by DSPC, which forms a more rigid bilayer at body temperature. The relative distribution of the liposome radiolabel was found to be similar when GM1-containing fluid vesicles (GM1-PC-C) are compared to DPPG-containing solid vesicles (DPPG-DSPC-C). However, with GM1-containing solid vesicles (GM1-DSPC-C) the distribution was favorably changed toward higher concentration in blood and decreased accumulation in liver-spleen. In this case, more than 50% of the recovered dose is found in carcass-skin and other organs, suggesting that liposomes can localize in significant amounts in tissues other than liver and spleen, provided that a long circulation time is achieved. Thus it appears from the comparison shown in Table 1 that liposome size, type of surface charge and bilayer fluidity all play a role in determining some of the optimal characteristics associated with prolonged circulation time of liposomes.

Using a formulation that gives results similar to the GM_1-DSPC-C, but composed of modified natural lipids from plants, we have examined the pharmacokinetics of doxorubicin (DXR), one of the most widely used anti-tumor drugs (Gabizon et al, 1989). Such liposome formulation composed of phosphatidylinositol and phosphatidylcholine (both hydrogenated from soy-beans) and cholesterol (HPI-HPC-C, 1:10:5 mole ratio), was loaded with DXR. Control liposomes were made of phosphatidylglycerol and phosphatidylcholine (both from egg yolks), and cholesterol (PG-PC-C), similarly loaded with DXR. The particle size of both liposome preparations was approximately 0.1 μm. The plasma pharmacokinetics and tissue distribution of DXR both free (soluble) and encapsulated in liposomes were examined after a single IV bolus dose of 10 mg/kg in BALB/c mice. Striking differences in various key pharmacokinetic parameters were observed, indicating an approximately 200-fold increase in the concentration of plasma DXR (liposome-encapsulated over free) integrated over a time period of 24 h (Gabizon et al, 1989).

TISSUE DISTRIBUTION IN TUMOR-BEARING MICE

Various liposome compositions were injected in tumor-bearing mice to examine their localization in tumors as compared to other tissues. The J6456 tumor, a T-cell-derived lymphoma (Gabizon and Trainin, 1980) was inoculated IM in the hind limb of syngeneic BALB/c mice. Liposomes were injected intravenously into mice with tumor implants weighing between 0.5 and 2 g. The results obtained 24 h after injection are presented in Table 2. A high significant increase in tumor uptake (up to 25-fold) was observed with liposome formulations selected for longer circulation times in normal mice. The highest tumor uptake (5.3% of injected dose/g) was obtained with GM1-DSPC-C liposomes. Another interesting finding is that liposome concentration in tumors has higher than that seen in all other tissues except for liver and spleen (data not shown for spleen). These values were obtained after correcting for the blood content of tumors of similar size as determined by Indium 111 oxine-labeled red blood cells (Heaton et al, 1979), indicating that the increased liposome concentration in tumors was not due to the tumor vascular blood pool. When free gallium-67-deferoxamine was injected, the uptake by tumor at 24 h was <0.1% of injected dose per gram. Other results not reported here in detail indicate that liposomes with long circulation times show a similarly enhanced accumulation in two other tumor models, mouse B16 melanoma, and human LS174 colon carcinoma. These data support the hypothesis that accumulation of liposomes in tumors depends on a prolonged circulation time. Since localization in other important tissues

Table 2. Tissue Distribution of Liposomes in Tumor-Bearing Mice[a]

Liposome Composition[b] (Molar ratio)	% Injected Dose/G (SD)		
	Tumor	Blood	Liver
PG-PC-C (1-9-5)	0.2 (0.0)	0.5 (0.2)	36.4 (6.3)
DSPC-C (10-5)	2.1 (0.3)	0.2 (0.0)	36.5 (7.6)
DPPG-DSPC-C-C (1-10-5)	4.1 (1.7)	0.3 (0.0)	38.3 (0.5)
HPI-DSPC-C (1-10-5)	4.1 (1.2)	1.4 (0.3)	37.8 (0.4)
GM1-PC-C (1-9-5)	3.5 (0.6)	2.6 (0.2)	20.7 (1.0)
GM1-DSPC-C (1-10-5)	5.3 (1.0)	3.1 (0.5)	31.7 (1.5)

[a]BALB/c female mice were inoculated IM with 10^6 J6456 tumor cells in the hind leg. Two to three weeks later, mice received each of the indicated formulations at a dose of 1 µmol phospholipid per mouse and were sacrificed 24 hours later. Average tumor weight ranged from 0.5 to 2.0 g. Total body recovery of radiolabeled material ranged from 50-70% of injected dose.
[b]All liposome preparations were extruded through a 0.05 0.1 µm polycarbonate membrane and had a particle size distribution of approximately 0.07-0.12 µm.

is not enhanced to the same degree, as suggested in this study, the new liposome formulations provide an opportunity for selective delivery of drugs to tumors.

DISCUSSION

Our formulation design was based on the following information regarding stability and blood half-life of liposomes and other particulate carriers. The inclusion of cholesterol (C) and phospholipids of high-temperature phase transition such as distearoylphosphatidylcholine (DSPC) has been shown to increase liposome stability in plasma as determined by the degree of retention of various liposome-encapsulated markers in vitro (Senior and Gregoriadis, 1982; Mayhew et al, 1979; Allen, 1981). Formulations containing sphingomyelin or DSPC are cleared more slowly after intravenous injection than those made with lipids of low phase transition temperature such as phosphatidylcholine (PC) from egg yolk (Gregoriadis and Senior, 1986; Hwang et al, 1980; Proffitt et al, 1983). The inclusion of certain gangliosides, conferring to the vesicle surface a negative charge and extra hydrophilicity, synergizes with cholesterol to enhance liposome stability in plasma (Allen et al, 1985) and prolongs liposome half-life in blood with a concomitant decrease in liver and spleen uptake (Allen and Chonn, 1987; Papahadjopoulos and Gabizon, 1987). Hydrophilic coating of polystyrene particles with polymeric glycols also results in significantly enhanced blood half-lives (Davis and Illum, 1986). Finally, it has been generally observed that reduction of liposome size contributes to slower clearance rate from the circulation (Senior, 1987).

The significance of the studies reported here is related to the possibility of increasing the concentration of anti-cancer agents in tumors. Our recent studies indicate that this can be achieved by using liposomes with reduced rate of RES-mediated clearance, thereby diminishing the likelihood of RES toxicity while simultaneously increasing the drug concentration in the vicinity of the tumor cells.

The accumulation of liposomes into the implanted tumors reported here, could be due to convective transport through leaky endothelia, as has been noted before for various tumors (Jain and Gerlowski, 1986). Such accumulation should be differentiated from the serendipitous targeting of most liposomes to liver and spleen which involves their recognition by the RES cells (Mayhew and Papahadjopoulos, 1983). Accumulation in tumors is apparently related to the ability of liposomes to stay in the circulation for longer periods of time, which in turn is controlled by the surface chemical characteristics that determine the extent of their recognition by the RES. Since we have no information that the observed accumulation in tumors depends on specific recognition, we define the phenomenon as non-specific targeting. This definition helps to differentiate it from ligand-specific targeting (Heath et al, 1980; Leserman et al, 1980; Huang et al, 1980). Such non-specific tumor targeting with liposomes may have the advantage of general applicability and simplicity compared with other approaches relying on recognition of specific antigens expressed by tumor cells, such as antibody-conjugated liposomes (Straubinger et al, 1988; Matthay et al, 1989), toxins and drugs (Davies and Crumpton, 1982) or radionuclides (Order, 1987).

Acknowledgements

We thank Nina Lopez, Renee Shiota and Ann Stubbe for technical assistance.

This work was supported until November, 1987 by a grant from the National Cancer Institute (CA 35340) and subsequently by Liposome Technology, Inc. (Menlo Park, CA).

REFERENCES

Allen, T.M., 1981, A study of phospholipid interactions between high density lipoproteins and small unilamellar vesicles, Biochim.Biophys.Acta, 640:385.

Allen, T.M. and Chonn, A., 1987, Large unilamellar liposomes with low uptake into the reticuloendothelial system, FEBS Lett., 223:42.

Allen, T.M., Ryan, J.L. and Papahadjopoulos, D., 1985, Gangliosides reduce leakage of aqueous-space markers from liposomes in the presence of plasma, Biochim.Biophys.Acta, 818:205.

Connor, J., Yatvin, M.B. and Huang, L., 1984, pH-sensitive liposomes: Acid-induced liposome fusion, Proc.Natl.Acad.Sci.USA, 81:1715.

Davies, A.J.S. and Crumpton, M.J., eds. 1982, in: "Experimental Approaches to Drug Targeting", Cancer Surveys 1(3):347.

Davis, S.S. and Ilum, L., 1986, Colloidal delivery systems: opportunities and challenges, in: "Site-Specific Drug Delivery", E. Tomlinson and S.S. Davis, eds., J. Wiley and Sons.

Duzgunes, N., Straubinger, R.M., Baldwin, P.A., Friend, D.S. and Papahadjopoulos, D., 1984, Proton-induced fusion of oleic acid/phosphatidylethanolamine liposomes, Biochemistry, 24:3091.

Ellens, H., Bentz, J. and Szoka, F., 1984, H+ and Ca2+-induced fusion and destabilization of liposomes, Biochemistry, 23:1532.

Gabizon, A. and Trainin, N., 1980, Enhancement of growth of a radiotion-induced lymphoma by T cells from normal mice, Br.J.Cancer, 42:551.

Gabizon, A. and Papahadjopoulos, D., 1988, Liposome formulations with prolonged circulation time in blood and enhanced uptake by tumors, Proc.Natl.Acad.Sci., 85:6949.

Gabizon, A., Huberty, J., Straubinger, R.M., Price, D.C. and Papahadjopoulos D., 1988, An improved method for in vivo tracing and imaging of liposomes using a gallium-67-deferoxamine complex, J.Liposome Res., 1:124.

Gabizon, A., Shiota, R. and Papahadjopoulos, D., 1989, Pharmacokinetics and tissue distribution of doxorubicin encapsulated in stable liposomes with long circulation times, J.Nat.Cancer Inst., 81:1485.

Gregoriadis, G. and Senior, J., 1986, Liposomes in vivo: a relationship between stability and clearance, in: "Targeting of Drugs with Synthetic Systems", G. Gregoriadis, J. Senior and G. Poste, eds., Plenum Press, New York.

Heath, T.D., Fraley, R. and Papahadjopoulos, D., 1980, Antibody targeting of liposomes, Science, 210:539.

Heaton, W.A., Davis, H.H., Welch, M.J., Mathias, C.J., Joist, H.H., Sherman, L.A. and Siegel, B.A., Br.J.Haematol., 42:613.

Huang, A., Huang, L. and Kennel, S.J., 1980, Monoclonal antibody covalently coupled to fatty acids: a reagent for in vitro liposome targeting, J.Biol.Chem., 255:8015.

Hwang, K.J., Luk, K.K. and Beaumier, P.L., 1980, Hepatic uptake and degradation of unilamellar sphingomyelin/cholesterol liposomes: a kinetic study, Proc.Natl.Acad.Sci.USA, 77:4030.

Jain, R.K. and Gerlowski, L.E., 1986, Extravascular transport in normal and tumor tissues, Crit.Rev.Oncol.Hematol., 5:115.

Leserman, L.D., Barbet, J., Kourilsky, F.M. and Weinstein, J.N., 1980, Targeting to cells of fluorescent liposomes covalently coupled with monoclonal antibody, Nature (London), 288:602.

Matthay, K.K., Abai, A., Cobb, S., Hong, K. and Papahadjopoulos, D., 1989, Role of ligand in antibody-directed endocytosis of liposomes by human T-leukemia cells, Cancer Res., 49:4879.

Mayhew, E. and Papahadjopoulos, D., 1983, Therapeutic applications of liposomes, in: "Liposomes", M.J. Ostro, ed., Marcel Dekker, New York, Inc.

Mayhew, E., Rustum, Y., Szoka, F. and Papahadjopoulos, D., 1979, Role of cholesterol in enhancing the antitumor activity of cytosine arabinoside entrapped in liposomes, Cancer Treat.Rep., 63:1923.

Order, S., 1987, ed., in: "Labeled and Unlabeled Antibody in Cancer Diagnosis and Therapy", NCl Monogr. 3.

Papahadjopoulos, D. and Gabizon, A., 1987, Targeting of liposomes to tumor cells in vivo, Ann.NY Acad.Sci., 504:64.

Papahadjopoulos, D., Jacobson, K., Nir, S. and Isac, T., 1973, Phase transitions in phospholipid vesicles: fluorescence polarization and permeability properties concerning the effect of temperature and cholesterol, Biochim.Biophys.Acta, 311:330.

Poste, G., 1983, Liposome targeting in vivo: problems and opportunities, Biol.Cell, 47:19.

Proffitt, R.T., Williams, L.E., Presant, C.A., Tin, G.W., Uliana, J.A., Gamble, R.C. and Baldeschwieler, J.D., 1983, Liposomal blockage of the reticulendothelial system: improved tumor imaging with small unilamellar vesicles, Science, 220:502.

Senior, J.H., 1987, Fate and behaviour of liposomes in vivo: a review of controlling factors, CRC Crit.Rev.Ther.Drug Carrier Systems, 3:123.

Senior, J. and Gregoriadis, G., 1982, Stability of small unilamellar liposomes in serum and clearance from the circulation: effect of phospholipid and cholesterol components, Life Sci., 30:2123.

Straubinger, R.M., Lopez, N.G., Debs, R.J. and Papahadjopoulos, D., 1988, Liposome-based therapy of human ovarian cancer: liposome-cell interaction control potency of negatively-charged and antibody-targeted liposomes, Cancer Res., 48:5237.

Weinstein, J.N., Magin, R.L., Yatvin, M.B. and Zaharko, D.S., 1979, Liposomes and local hyperthermia: selective delivery of methotrexate to heated tumors, Science, 204:188.

Yatvin, M.B., Weinstein, J.N., Dennis, W.H. and Blumenthal, 1978, Design of
    liposomes for enhanced local release of drugs by hyperthermia,
    Science, 202:1290.
Yatvin, M.B., Kreutz, W., Horwitz, B.A. and Shinitzky, M., 1980, pH-
    sensitive liposomes: possible clinical implications, Science,
    210:1253.

STABILIZATION OF LIPID MICROSTRUCTURES: FUNDAMENTALS AND APPLICATIONS

Alan S. Rudolph, Alok Singh, Ronald R. Price, Beth Goins and
Bruce P. Gaber

Center for Bio/Molecular Science and Engineering, Code 6090
Naval Research Laboratory, Washington, DC 20375-5000, USA

INTRODUCTION

The thermotropic and lyotropic phase behavior of components that comprise lipid assemblies such as liposomes results in an inherent instability of these structures when exposed to extremes of temperature and hydration. This can present significant limitations to their successful application. As liposomes and other macroassemblies of lipid molecules progress toward application, considerable efforts have been made to improve the stability of these structures. We can define stabilization of lipid microstructures as the ability to withstand chemical, mechanical, or thermal extremes which may be encountered in the variety of applications that are being pursued. In particular, the definition of stabilization for drug delivery and slow release purposes should include increased persistence in the body and avoidance of the reticular endothelial system (RES) which will result in enhanced activity of encapsulants in vivo.

We shall describe several stabilizing strategies which our group has pursued. One such strategy is the stabilization of lipid microstructures to reduced water states such as freezing and lyophilization through the addition of small molecular weight polyhydroxy compounds such as the disaccharides sucrose and trehalose. The molecular action of these water replacement molecules in altering the phase behavior of phospolipids has been recently elucidated (Crowe et al, 1988; Rudolph et al, 1986). We have employed these stabilizing molecules in liposome systems of interest to our group such as the blood surrogate liposome encapsulated hemoglobin (LEH) and have documented the usefulness of these compounds for long-term dry storage of LEH (Rudolph, 1988).

The synthetic modification of lipids to include polymerizable moieties such as diacetylenes, olefins, or methacrylates is another approach toward stabilization of lipid microstructures. The polymerizable functional group is introduced into the acyl chains of phospholipids which upon polymerization results in a covalently linked hydrophobic region in the bilayer. Polymerization can be induced by photons, radical initiators, or redox polymerization (O'Brien et al, 1981; Kusumi et al, 1983). Depending on the degree of polymerization achieved, thermotropic changes in the bilayer may be inhibited which may result in alterations in bilayer permeability and mechanical stability (O'Brien et al, 1981; Kusumi, 1983). We have also discovered that the inclusion of polymerizable groups plays a role in determining the morphology of the assembly. An example of this is the hollow cylindrical structures (or tubules) that form from the long chain diacetylenic lecithins (Yager and Schoen, 1984). The physicochemical properties of these structures have been investigated and stabilization strategies developed for their use in biomaterial applications such as a slow release vehicle for antifouling agents.

STABILIZATION OF LIPID MICROSTRUCTURES TO FREEZE/THAW AND FREEZE-DRYING BY THE ADDITION OF SMALL MOLECULAR WEIGHT WATER REPLACEMENT MOLECULES

The stabilization of liposomes to low temperature and reduced water states has centered on the addition of low molecular weight polyols such as carbohydrates and certain amino acids (Crowe et al, 1988; Rudolph et al, 1986). The particular solutes which have proved most effective have been those which accumulate in organisms that experience freezing and desiccating environments. Thus, the fields of cryobiology and anhydrobiosis have added much to our knowledge of useful agents for liposome applications.

The driving force for self-assembly of phospholipids into liposomes and other microstructures is the amphiphilic nature of these molecules and the energetic cost (reflected in an entropic component) of exposing hydrophobic regions to the polar solvent water (Tanford, 1981). Thus, the inherent stability of a bilayer assembly is compromised by the removal of water that occurs during drying conditions. In more complex membranes, the reduction in water content can result in phase separation of different classes of lipids and the loss of membrane protein activity (Crowe et al, 1983). Phase separation may also result in compromised permeability as phase boundaries are thought to be areas of increased permeability (Crowe et al,1983). Extensive freezing may also result in dehydration although the decreased temperature can induce phase separation based on the thermotropic phase behavior of the membrane components.

The action of small molecular weight solutes has recently focused on their direct interaction with lipids and lipid asemblies. It has been hypothesized, on the basis of calorimetric and infrared spectroscopic studies, that the disaccharides hydrogen bond to the interfacial region of the bilayer (Crowe et al, 1985). Based on the experimental evidence, a molecular model of trehalose-DMPC has been constructed and subjected to energy minimization (Rudolph et al, 1990) (Fig. 1). It is thought that the interaction of trehalose (and other disaccharides) and phospholipids results in the maintenance of the bilayer in a semi-fluid state even in the absence of water (Lee et al, 1986; Lee et al, 1989). In practice, the result of this action is to prevent vesicle fusion and loss of entrapped contents during freeze-drying (Crowe et al, 1986).

We have successfully employed trehalose to stabilize liposome encapsulated hemoglobin (LEH) in the dry state (Rudolph, 1988). LEH is comprised of distearoyl phosphatidylcholine, cholesterol, dimyristoyl phosphatidylcholine and alpha tocopherol (in a 10:9:0.9:0.1 mole ratio). The procedure for including the protective solutes in the production of LEH has been to add the solutes to the hemoglobin before hydration of the dry lipid mixture. The resultant multilamellar suspension is then processed using a high pressure hydrodynamic shear apparatus which creates unilamellar vesicles 0.2-0.5 microns in diameter (Beissinger et al, 1986). Successive washing of LEH with trehalose-phosphate buffered saline to remove unencapsulated hemoglobin can be accomplished which results in trehalose inside LEH and in the extravesicular compartment. We have found the optimal amount of trehalose in the preservation of LEH to be 150-300 mM (Rudolph, 1988). The measure of stability in these studies is the ability of LEH to retain hemoglobin and the maintenance of vesicle size. Figure 2 shows the effect of trehalose on

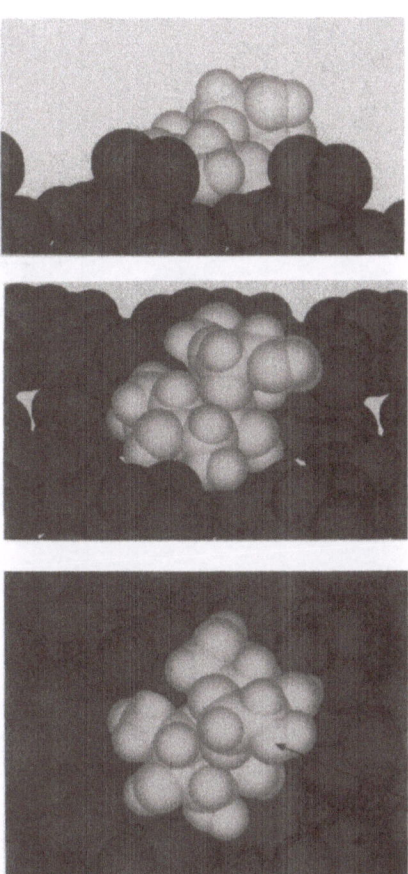

Fig. 1. Molecular model depicting the interaction of trehalose with a monolayer of DMPC. Models of this type provide the starting point for energy minimization studies of the interaction of trehalose, sucrose, and glucose with the lipid bilayer. The model was realized using the graphics program NanoVision™ on the Apple Macintosh personal computer.

maintaining LEH size following rehydration. Unprotected vesicles undergo extensive vesicle fusion as evidenced by the large increase (from 0.3 to 1.2 microns) in the hydrodynamic diameter measured by quasi-elastic light scattering. In contrast, LEH produced with 150–300 mM trehalose maintains vesicle size following rehydration at any time point examined. The inhibition of LEH fusion by trehalose is also manifested by the retention of hemoglobin following rehydration (Fig. 3). Unprotected vesicles lose 70% of the encapsulated hemoglobin following rehydration, while those vesicles containing 150–300 mM trehalose retain 80% of their encapsulated hemoglobin following rehydration at three months. These results demonstrate that trehalose is able to preserve LEH for at least three months in the dry state. We are currently examining samples preserved for longer periods of time.

In addition to the ability of these agents to stabilize lipid bilayers, it appears that their action also is observed in the preservation of proteins in the dry state (Carpenter et al, 1988). For LEH, this is observed in the stabilization of hemoglobin from oxidative conversion to methemoglobin (which does not carry oxygen) by the addition of trehalose (Fig. 4). Unprotected vesicles experience a large conversion of hemoglobin to methemo-

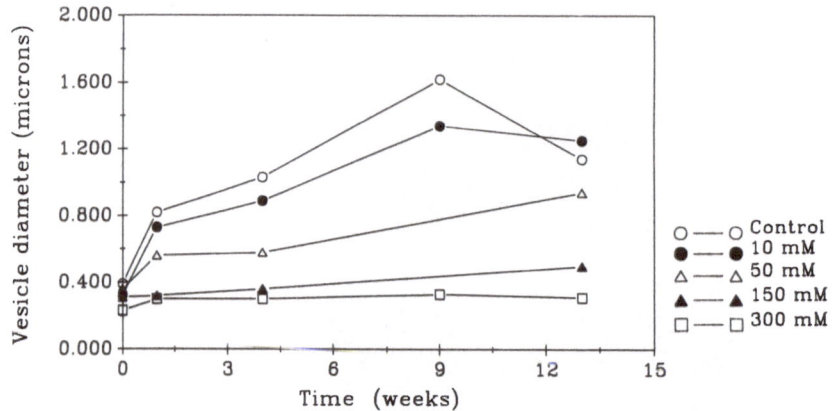

Fig. 2. Photon correlation spectroscopic results of vesicle size following rehydration after dry storage (in vacuum) for varying lengths of time. Vesicle diameter is determined from a best fit analysis of the correlation function.

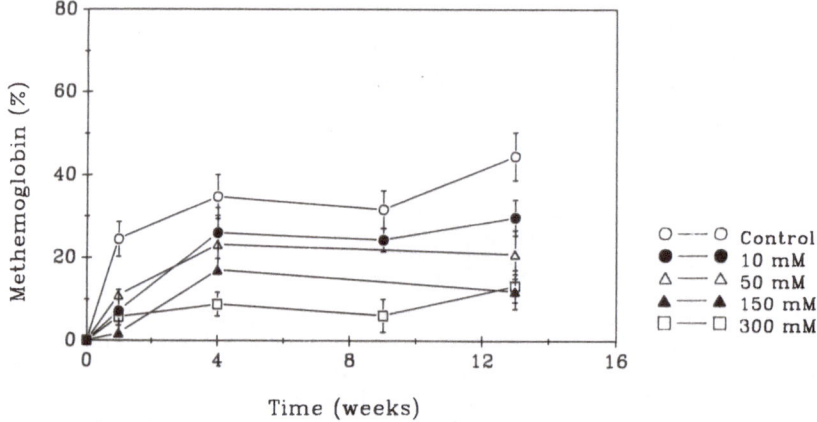

Fig. 3. Retention of intravesicular hemoglobin following rehydration from dry storage for varying lengths of time.

globin. LEH samples with increasing trehalose have less accumulated methemoglobin, the best preserved samples at three months experiencing a 10% increase in methemoglobin compared to 40-50% in the untreated samples.

The application of protective solutes to preserve the function and structure of lipid microstructures can be extended to other microstructures in addition to liposomes. We have reported the successful use of the disaccharide trehalose in preserving the structure of tubules (formed from diacetylenic lecithins) following freeze-drying (Rudolph et al, 1988a). The use of protective solutes to stabilize other assemblies such as microemulsions or micelles has not been investigated.

STABILIZATION OF LIPID MICROSTRUCTURES BY CHEMICAL MODIFICATION OF LIPIDS TO INCLUDE POLYMERIZABLE GROUPS

One strategy which we have pursued to stabilize lipid microstructures is the chemical modification of lipids by synthetic methods to include

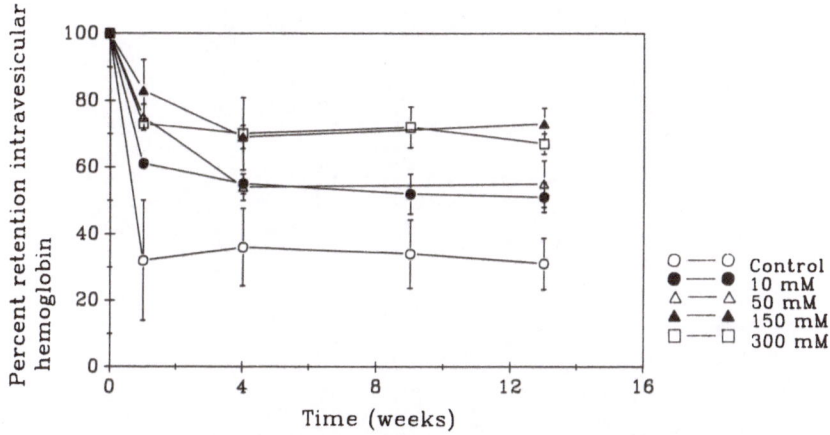

Fig. 4. Methemoglobin levels (expressed as % increase over control) of rehydrated LEH that has been stored dry in vacuum for varying lengths of time.

polymerizable moieties. Polymerization of the resultant microstructures formed from the modified lipid creates a morphology which may be mechanically enhanced conferring resistance to extremes in temperature or solvent conditions.

Our focus has been to modify the acyl chain region of phospholipids and to create other two-chain amphiphiles that contain polymerizable groups in the hydrophobic region. Perhaps the most studied class of lipids in our hands has been the diacetylenic lecithins (Singh and Schnur, 1985) (Fig. 5). The inclusion of the diacetylenic groups gave us the opportunity to investigate the self-assembling properties, subsequent polymerization mechanisms, and the effect of polymerization on microstructure morphology. Polymerization of the diacetylenes proceeds upon uv exposure at 254 nm, by a 1,4 addition reaction to result in the colored conjugated polymer backbone (Singh and Schnur, 1985). To date, characterization of the polymer for molecular weight distribution and conjugation length have not been accomplished. We have, however, observed that the conjugated polymer backbone

$$CH_2-O-\underset{O}{\overset{\parallel}{C}}-(CH_2)\overline{_m}C\equiv C-(CH_2)\overline{_p}C\equiv C-(CH_2)\overline{_n}R$$

$$CH-O-\underset{O}{\overset{\parallel}{C}}-(CH_2)\overline{_m}C\equiv C-(CH_2)\overline{_p}C\equiv C-(CH_2)\overline{_n}R$$

$$CH_2-O-\underset{O_-}{\overset{O}{\overset{\parallel}{P}}}-O-CH_2-CH_2-\overset{+}{N}Me_3$$

1a.	m= 5 - 15	n= 6 - 16	p= 0	R= CH_3
1b.	m= 4, 8	n= 13, 9	p= 1	R= CH_3
1c.	m= 8	n= 8	p= 0	R= $CH=CH_2$

Fig. 5. Diacetylenic phosphatidylcholines. We have synthesized diacetylenic phosphatidylcholines with varying methylene segments above the diacetylene (m), between the diacetylenes (p), and below the diacetylenes (n). Most of these compounds exhibit the formation of tubules in their phase behavior. Compounds made are listed as 1a-1c.

107

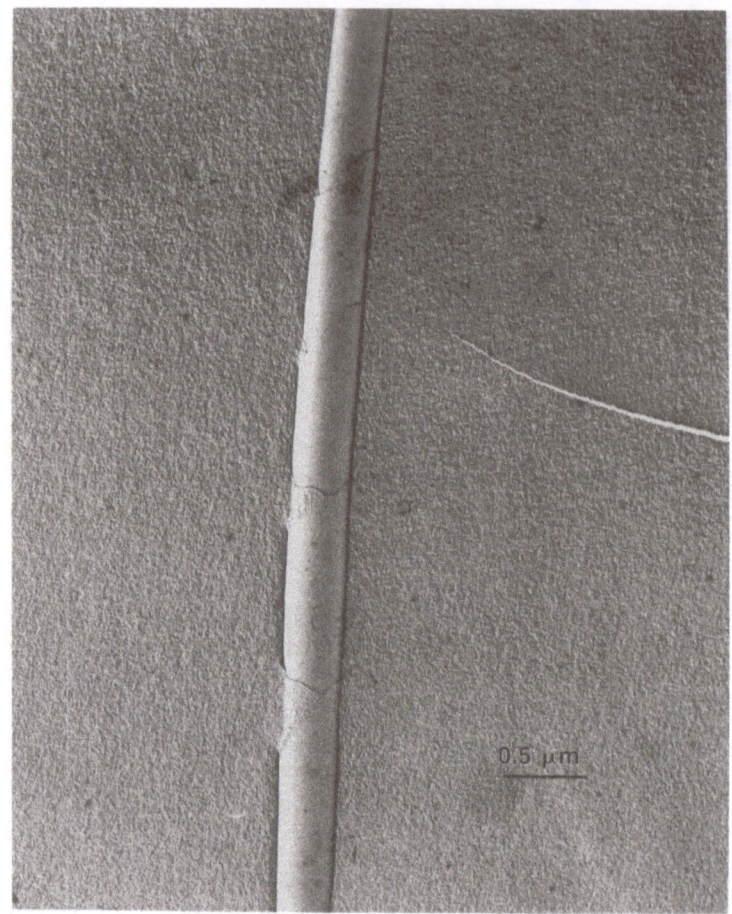

Fig. 6. Freeze-fracture electron micrograph of a tubule formed from a
diacetylenic lecithin. Note the helical wrapping of bilayers along
the length of the tubule. Tubule walls usually consist of a number
of diacetylenic bilayers.

present in the lipid bilayer results in a reversible thermochromic effect in
polymerized vesicles (Singh et al, 1986). This suggests that the polymer
characteristics can be modulated by the thermotropic nature of the lipid
bilayer. In addition, we have examined potentially useful applications of
polymerizable microstructures such as polymerizable vesicles that contain
bacteriorhodopsin (Ahl et al, 1989) or films and flat bilayers that incorporate acetylcholine receptors (Ligler et al, 1988). These are fundamental
steps toward the use of stabilized microstructures in biosensor applications.

During the course of our study of the diacetylenic lecithins, it was
discovered that these lipids have unusual self-assembling properties. Yager
and Schoen (1984) first observed the formation of long, hollow cylindrical
structures, or tubules (Fig. 6). We have learned much about the formation
of these structures, although the driving forces that lead to the tubule
structure have not been elucidated clearly. Most of the diacetylenic
lecithins we have made, with the exception of methylene interrupted diacetylenes, form tubules. This may suggest that the chain conformation
assumed by the presence of the diacetylene is somehow important in the
ability of these lipids to form tubules. We have also shown that chirality

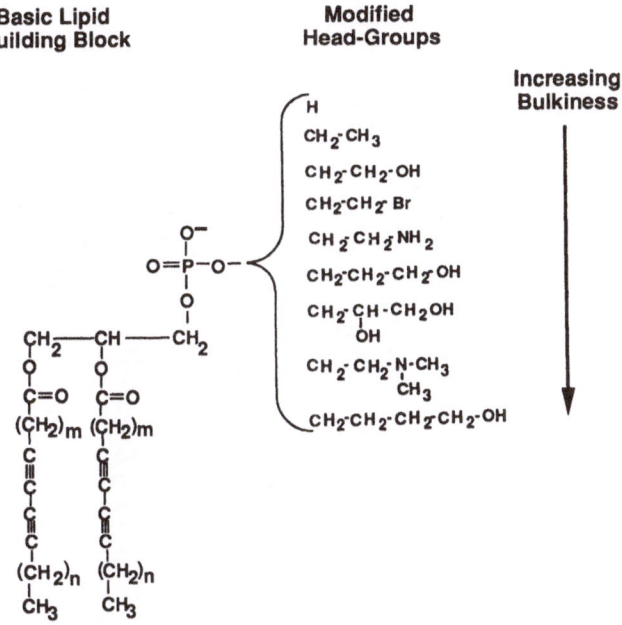

Fig. 7. Synthetic phosphatide family of head-group modified diacetylenes. The compounds represent head groups with increasing bulkiness and differing charge. These compounds have been used to examine the role of the interfacial region in the self-assembly of lipid micro-structures.

is not an absolute requirement for an amphiphile to form tubules as we have made tubule-forming ammonium surfactants that contain diacetylenes without a chiral center (Singh and Schnur, 1988). The experimental evidence to date, which includes calorimetric (Burke et al, 1988) raman and infrared spectro-scopic (Schoen and Yager, 1985; Rudolph and Burke, 1987; Rudolph et al, 1986), and x-ray (Rhodes et al, 1988) studies suggest that the tubules formed from the diacetylenic lecithins are highly ordered structures that form from bilayers that have reduced curvature (Burke et al, 1988; Rudolph and Burke, 1987; Rudolph et al, 1988b).

We are currently examining modification of the head group region of the polymerizable lecithin molecule to determine the role of the interfacial forces on the formation of microstructure morphology. A representative list of the compounds we have made is presented in Fig. 7. Considerations we are examining with these molecules are the role of charge and head group size in the self-assembling properties of these phospholipids. The modification of this region to form diacetylenic phosphatidylhydroxyethanol (Fig. 7) has resulted in the formation of new microstructures which appear as rope-like, twisted helical structures (Fig. 8). We are currently examining the parameters surrounding the formation of these structures.

The formation of tubules by the diacetylenic lecithins makes their use for liposomal applications difficult as the liposomal structure in water is lost below the calorimetric transition observed at 39°C. This problem can be alleviated by stabilizing liposomal structures containing diacetylenes. This is accomplished by mixing the diacetylene with a small mole percent of fully saturated lecithins (Singh and Gaber, 1988). The perturbation of the crystallinity of the chains below the phase transition may be responsible for the inhibition of tubule forming properties. The insertion of the fully saturated lecithins into bilayers of diacetylenic lecithins results in the

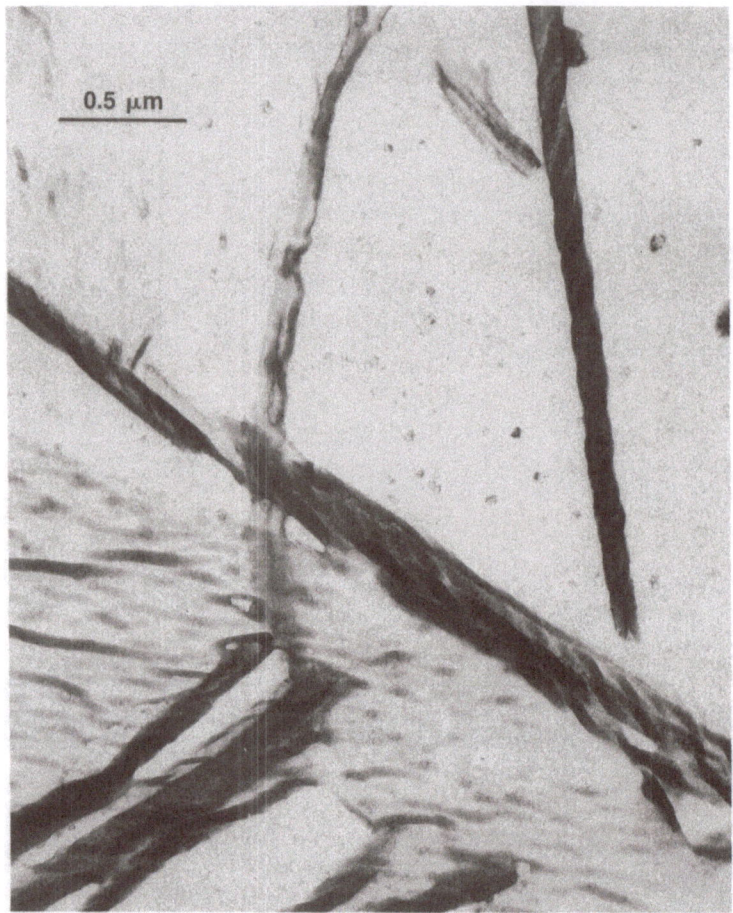

Fig. 8. Negative stain electron micrograph of the tubule-like microstructures formed from bis-1,2(10,12)-tricosadiynoyl-sn-glycero-3-
phosphatidylhydroxyethanol. These structures have tighter wrapping
than the tubules observed from the diacetylenic lecithins.

maintenance of liposomal structure below the thermotropic phase transition
of the lipid mixture. This strategy has been used to incorporate bacteriorhodopsin in liposomes containing diacetylenic lecithins (Ahl et al, 1989).
In addition, the mixing of the saturated lecithins with the diacetylenic
lecithin appears to change the conjugation length of the resultant polymer
with the possibility of longer conjugation lengths achieved.

    We have also mixed diacetylenes with other polymerizable groups to
create heterobifunctional polymerizable lecithins (Singh et al, 1986). In
these synthetic schemes vinyl or olefin groups have been added to the terminal end of the acyl chains. The hydrophobic region then contains two
potential sites for polymerization. One group, such as the diacetylene may
be important in the formation of a particular morphology such as the tubule,
while the other provides the principal site for polymerization.

We have investigated the use of tubules for controlled release applications. Conceptually, the incorporation of materials within a tubule could involve the encapsulation of aqueous material between the lamellae of tubule walls, the lumen of the tubule (usually 0.3-0.7 microns in diameter), or intercalated in the bilayers of the tubule. Early methods of encapsulation demonstrated that trapped liposomes in the tubule structure could be used as slow release carriers (Burke et al, 1987). For some applications, an important step toward the use of tubules in slow release applications has been the development of electroless plating techniques to coat the surface of the tubule with a thin layer of metal (Rudolph et al, 1989). This is accomplished by first treating the tubule surface with a colloidal catalyst. The metal is subsequently deposited onto the surface of the tubule (Fig. 9). We have accomplished this with a number of metals such as Ni, Cu, Au, Ag, Pd. In these applications the material encapsulated into the tubule becomes encased in a metal sheath. One of the applications we have examined using this technology is the encapsulation of antifouling agents within the tubule to include in a paint that can be applied to the bottom of ship hulls.

The fouling layer that accumulates on the ship hull increases drag on the hull and decreases performance of the vessel while increasing energy costs. Thus the control of fouling organisms is of prime importance to the efficient operation of the vessel. Currently, the bottom paint formulations contain large loadings of highly toxic metals such as tributyl tin compounds which have come under close governmental regulation. Many paint producers have returned to very high loading copper containing paints to meet modern

Fig. 9.   Scanning electron micrograph of copper coated tubules. Note the open lumen of two tubules in the center of the micrograph. Other metals have been deposited on tubules using electroless plating techniques.

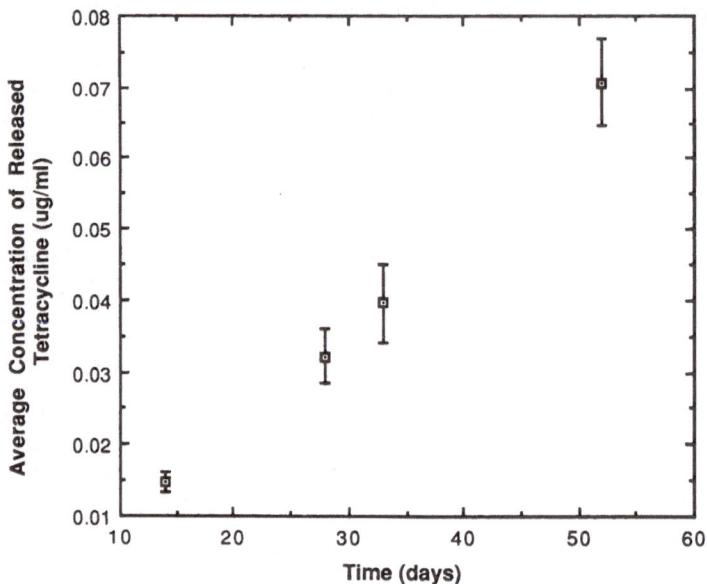

Fig. 10. Concentration of released tetracycline found in seawater bathing slides painted with tubule-based bottom paint over the course of nearly three months. Panels painted with tetracycline-loaded copper tubule paint and submerged in the Chesapeake Bay show reduced fouling compared to commercial antifouling paints.

ship service cycle requirements. Despite the very heavy loading of copper in the paint these paints have been shown to fail in as little as nine months of use, with average lifetimes of 14-18 months. The cost in labor and materials of a limited lifetime bottom paint is high. We have investigated the use of tubules as a means to slowly release antifouling agents and reduce the recruitment of marine organisms to the ship hull surface. In addition, the use of slow release encapsulants may increase the lifetime and ultimate economy of antifouling bottom paints.

Tubules formed from lipids have proven to be easily metalized with copper and we have been able to encapsulate secondary materials such as tetracycline, benzoic acid, and ammonium surfactants all of which have been shown to enhance the performance of copper toxicants. The tubules form a short fiber composite with the base coating resin and thus resist mechanical erosion better than a solid particulate. Further, it has been shown that the tubules are able to control the release rate of tetracycline within the antifouling paint effectively from coatings which are free from leaching rosin. Following a nearly three month dwell period, samples of artificial seawater which were exposed to the tubule-loaded coatings released tetracycline at a rate of .001 to .002 ug/ml/day from the tubule-loaded coating (Fig. 10). Environmental test panels that contained the tubule paint were exposed for a nine month period of time and showed minimal fouling. The same achievement with standard commercial paints requires six times the copper toxicant loading.

It is possible that the release mechanism is due to formations of large aggregates of interconnected tubules in the coating which might provide a network of channels permitting the intrusion of seawater into the coating. This would then function in lieu of chemical dissolution of rosin in a coating to release toxicants far within antifouling paint. By providing a pathway such that the intrusion of seawater would follow the path of dissolving materials within the tubule, water should only be able to penetrate a limited way into the coating based upon the amount of material released, not on mechanical or chemical dissolution of a majority component of the coating itself. In addition the use of tubular microstructures themselves may provide a greater flexibility in the use of a range of toxicant and repellent materials either alone or in a synergistic combination to increase the antifouling performance of the coating as technology into natural based toxicants and repellents progresses.

SUMMARY

We are continuing to explore two strategies in the stabilization of lipid microstructures; (1) the addition of stabilizing molecules to microstructure formulations such as water replacement molecules (e.g. trehalose) and (2) the chemical modification of lipids to include stabilizing functional moieties. Both of these approaches rely on a strong fundamental understanding of the principals of self-assembly and thermotropic and lyotropic phase behavior of the assembled microstructure. For some applications (such as the antifouling paint), the lipid microstructures form useful templates for the creation of structurally enhanced biomaterials such as metal coated tubules or liposomes. Our plan is to continue the study of stabilizing strategies for biological self-assemblies and add to both our fundamental understanding of this process and to useful applications which may arise from this endeavor.

Acknowledgements

We would like to thank Mr. Richard Cliff and Mrs. Marcia Patchan for their technical assistance and the Office of Naval Research, Naval Medical Research and Development Command, and the Defense Advance Research Projects Agency for their financial support.

REFERENCES

Ahl, P., Singh, A., Price, R., Smuda, J. and Gaber, B.P., 1989, Insertion of bacteriorhodopsin into polymerized diacetylenic phosphatidylcholine bilayers, Biophys.J., 55:321a.

Beissinger, R.L., Farmer, M.C. and Gossage, J.L., 1986, Liposome encapsulated hemoglobin as a red cell surrogate: preparation scale-up, Trans.Am.Soc.Artif.Org., 32:58.

Burke, T.G., Singh, A. and Yager, P., 1987, Entrapment of 6-carboxy fluorescein within cylindrical phospholipid microstructures, Ann.N.Y.Acad. Sci.USA, 507:330.

Burke, T.G., Rudolph, A.S., Sheridan, J.P., Dalziel, A., Singh, A. and Schoen, P.E., 1988, Calorimetric study of novel phase behavior of a phosphatidylcholine containing diacetylenes, Chem.Phys.Lipids, 48:215.

Carpenter, J.F., Crowe, J.H. and Crowe, L.M., 1988, Stabilization of phosphofructokinase with sugars during freeze-drying, Biochem.Biophys. Acta, 923:109.

Crowe, J.H., Crowe, L.M. and Jackson, S.A., 1983, Preservation of structure and functional activity in lyophilized sarcoplasmic reticulum, Arch.Biochem.Biophys. 220:477.

Crowe, J.H., Crowe, L.M. and Chapman, D., 1985, Interaction of carbohydrates with dry dipalmitoyl phosphatidylcholine, Arch.Biochem.Biophys., 236:289.

Crowe, L.M., Womersley, C., Crowe, J.H., Reid, D., Appel, L. and Rudolph, A.S., 1986, Prevention of fusion and leakage in freeze-dried liposomes by carbohydrates, Biochim.Biophys.Acta, 861:131.

Crowe, J.H., Crowe, L.M., Carpenter, J.F., Rudolph, A.S., Winstrom, C.A., Spargo, B. and Anchordougy, T.J., 1988, Interactions of sugars with membranes, Biochim.Biophys.Acta, 947:367.

Kusumi, A., Singh, M., Tirrell, D.A., Oehme, G., Singh, A., Samuel, N.K.P., Hyde, J.S. and Regen, S.L., 1983, Dynamic and structural properties of polymerized phosphatidylcholine vesicles membranes, J.Amer.Chem. Soc., 105:2975.

Lee, C.W.B., Waugh, J.S. and Griffin, R.G., 1986, Solid-state NMR study of trehalose/1,2-dipalmitoyl-sn-phosphatidylcholine interactions, Biochem., 25:3737.

Lee, C.B.W., Das Gupta, S.K., Mattai, J., Shipley, G.G., Abdel-Mageed, A., Makriyannis, A. and Griffin, R.G., 1989, Characterization of the L phase in trehalose stabilized dry membranes by solid-state NMR and x-ray diffraction, Biochem., 28:5000.

Ligler, F.S., Fare, T.L., Seib, K.D., Smuda, J.W., Singh, A., Ayers, M.E., Dalziel, A. and Yager, P., 1988, Fabrication of key components of receptor based biosensor, Med.Inst., 22:247.

O'Brien, D.F., Whitesides, T.H. and Klingbiel, R.T., 1981, The photopolymerization of lipid-diacetylenes in biomolecular-layer membranes, J.Polym.Sci: Polym.Letts.Ed., 19:95.

Rhodes, D.G., Blechner, S.L., Yager, P. and Schoen, P.E., 1988, Structure of polymerizable lipid bilayers. I- 1,2-bis(10,12-tricosadiynoyl)-sn-glycero-3-phosphocholine, a tubule forming phosphatidylcholine, Chem.Phys.Lipids, 49:39.

Rudolph, A.S., 1988, The freeze-dried preservation of liposome encapsulated hemoglobin: a potential blood substitute, Cryobiology, 25:277.

Rudolph, A.S. and Burke, T.G., 1987, A Fourier-transform infrared spectroscopic study of the polymorphic phase behavior of 1,2 bis(tricosa-10,12-diynoyl)-sn-3-phosphocholine; a polymerizable lipid which forms novel microstructures, Biochim.Biophys.Acta, 902:345.

Rudolph, A.S., Crowe, L.M. and Crowe, J.H., 1986, The effects of three stabilizing agents: proline, betaine, and trehalose on membrane phospholipids, Arch.Biochem.Biophys., 245:134.

Rudolph, A.S., Schnur, J.M., Singh, A., 1988a, The stabilization of a polymerizable lecithin with carbohydrates, Biophys.J., 53:120a.

Rudolph, A.S., Singh, B.P., Singh, A. and Burke, T.G., 1988b, Phase characteristics of positional isomers of 1,2-bis heptacosadiynoyl-sn-glycero-3-phosphocholine; tubule forming phosphatidylcholines, Biochim.Biophys.Acta, 943:454.

Rudolph, B., Chandrashekar, I., Gaber, B.P. and Nagumo, M., 1990, Molecular modelling of saccharide-lipid interactions, Chem.Phys.Lipids, in press.

Rudolph, A.S., Calvert, J.M., Schoen, P.E. and Schnur, J.M., 1989, Technological development of lipid based tubule microstructures, in: "Technological Applications of Lipid Microstructures", Advances in Experimental Medicine and Biology Series, vol. 238, B.P. Gaber, J.M. Schnur and D. Chapman, eds., Plenum Press, N.Y.

Schoen, P.E. and Yager, P., 1985, Spectroscopic studies of polymerized surfactants: 1,2-bis(10,12-tricosadiynoyl)-sn glycero-3-phosphocholine, J.Polym.Sci:Polym.Phys.Ed., 23:2203.

Singh, A. and Schnur, J.M., 1985, Polymerized diacetylenic phosphatidylcholine vesicles: synthesis and characterization, Polymer Preprints, 26(2):184.

Singh, A. and Schnur, J.M., 1988, Self-assembled microstructures from a polymerizable ammonium surfactant: Di (Hexacosa-12,14-Diynyl) dimethylammonium bromide, J.Chem.Soc.,Chem.Comm., 1222-1223.

Singh, A. and Gaber, B.P., 1988, Influence of short chain lipid spacers on the properties of diacetylenic phosphatidylcholine bilayers, in: "Applied Bioactive Polymeric Materials", C.G. Gebelein, C.E. Carraher, Jr. and V.R. Foster, eds., Plenum Press, N.Y.

Singh, A., Thompson, R.B. and Schnur, J.M., 1986, Reversible thermochromism in photopolymerized phosphatidylcholine vesicles, <u>J.Amer.Chem.Soc.</u>, 108:2785.

Singh, A., Price, R., Schoen, P.E. Yager, P. and Schnur, J.M., 1986, Tubule formation by heterobifunctional polymerizable lipids: synthesis and characterization, <u>Polymer Preprints</u>, 27(2):393.

Tanford, C., 1981, "The Hydrophobic Effect", John Wiley and Sons, N.Y.

Yager, P. and Schoen, P.E., 1984, Formation of tubules by a polymerizable surfactant, <u>Mol.Cryst.Liq.Cryst.</u>, 106:371.

NON-IONIC SURFACTANT VESICLES AS CARRIERS OF DOXORUBICIN

A.T. Florence*, C. Cable**, J. Cassidy[#] and S.B. Kaye[#]

*The Centre for Drug Delivery Research, The School of
Pharmacy, University of London, **The Department of Pharmacy
University of Strathclyde, Glasgow, UK and [#]Department of
Medical Oncology, University of Glasgow, UK

INTRODUCTION

Unilamellar or multilamellar vesicles prepared from synthetic non-ionic surfactants can be considered to be synthetic surfactant analogues of liposomes (Baillie, 1988; Florence and Baillie, 1989). The exploration of alternatives to phospholipids as the main lipid moiety in the fabrication of vesicles is in itself worthwhile, particularly as phospholipid raw materials are naturally occurring substances, not always in plentiful supply, and may be unstable during storage and processing. Methods for the processing of lipid raw material, for synthesizing, modifying and purifying natural lipids, are important for the ultimate large-scale clinical use of phospholipid-based vesicular delivery. Moreover, by studying the behaviour of vesicles constructed from nonionic surfactants (or other lipid species) there is the potential to determine what are the key chemical and physical factors which control biological distribution and targeting.

In considering alternatives to liposomes, the principal question is to what extent the phospholipid construction is essential for the reputed biological effects of liposomal delivery systems. For example, the so-called "membrane mimetic" nature of the phospholipid bilayer can, it is claimed, promote the cellular uptake of drugs which otherwise would not readily cross biological membranes. Obviously there is a wide variety of vesicle-forming phospholipids and not all have equal membrane activity. While there will be some specificity, the fact that non-ionic surfactants and other amphipathic materials form bilayer structures, and interact with cholesterol to form osmotically active analogues of liposomes, suggests that the physical chemistry rather than the chemical nature of the system is important. However, Burkhanov and coworkers (1988) have referred to the "unique surface topography" that may be vital in vivo. If surface charge is crucial then an obvious difference is that the non-ionic vesicles are only weakly charged through adsorption of hydroxyl ions to their hydrophilic head groups, unless the system contains a charged amphipath such as dicetyl phosphate (Azmin et al, 1985) or stearylamine, or indeed, ionic drugs such as doxorubicin or sodium stibogluconate which may also confer charge on the system (Stafford et al, 1988). The adsorption of serum proteins such as albumin onto the vesicular surface following injection leads to a change in the properties of the vesicle, particularly its permeability. There is some in vitro evidence that surfactant III NSVs containing 50 mol % cholesterol bind more protein

Targeting of Drugs, Edited by G. Gregoriadis *et al.*
Plenum Press, New York, 1990

than dipalmitoyl phosphatidylcholine (DPPC) liposomes with the same proportion of cholesterol. Uptake by <u>Tetrahymena ellioti</u> of surfactant I vesicles
and liposomes of DPPC, both with 50% cholesterol, is similar, so it is
likely that the chemical structure is important only in so far as it
determines the pattern of protein adsorption, the coated particle rather
than the native vesicle being recognized.

Unilamellar and multilamellar vesicles form from a wide variety of non-
ionic surfactants, such as alkyl glucosides, alkyl-, alkyl-aryl and cholesteryl polyoxyethylene ethers (Patel et al, 1984), steroidal lariat ethers
(Echegoyan et al, 1988) and alkyl and dialkyl glycerol ethers (Vanlerberghe
et al, 1972; 1978; Ribier et al, 1984; Azmin et al, 1985; Baillie et al,
1985). The list of vesicle-forming non-ionic substances is more extensive
than it once was, although the constraints of surfactant monomer geometry
and solubility still stand. It is the polyglycerol ethers which we have
studied for some time, particularly with regard to the delivery and efficacy
of anticancer drugs in animal systems (Azmin et al, 1985, 1986; Rogerson et
al, 1988; Kerr et al, 1988; Cable et al, 1988) while there has also been a
concerted effort in their use in leishmaniasis (Baillie et al, 1986; Carter
et al, 1988; Hunter et al, 1988).

As with liposomes, reliance is placed on the slow release of the drug
from the unilamellar or multilamellar system to achieve sustained activity,
or reduced toxicity, but targeting to organs other than those of the
reticuloendothelial system (RES) will be achieved only if extravasation of
the vesicular dispersions can take place. Targeting to specific cells (such
as macrophage cells) occurs by more specialized uptake processes, which
might be expected to depend on the chemistry of the vesicle (Burkhanov et
al, 1988). The extent to which specific organ and cellular targeting are
dependent on the chemical nature of the vesicle will determine the success
of these alternative materials as the basis of vesicular carriers. So far
the evidence is that non-ionic surfactant vesicles (niosomes or NSVs), and
liposomes display few differences <u>in vivo</u>; the evidence for extravasation is
scant, although in some studies described here, increased levels of drug
have been found in tumours. This, however, may be the result of binding of
vesicles to vessel walls. High analytical levels of drug do not necessarily
mean high levels of free, diffusible drug.

NON-IONIC SURFACTANTS AS THE BASIS OF VESICULAR CARRIERS

Although several classes of non-ionic surfactant have been shown to
form vesicles (Table 1), there is limited published work on <u>in vivo</u> studies
conducted with vesicles prepared from non-ionic surfactants. Our work has
concentrated on two surfactants, a hexadecyl triglycerol ether (Surfactant I)
and a dialkyl (C_{16}/C_{12}) heptaglycerol derivative (Surfactant II),
(Vanlerberghe et al, 1972, 1978; Handjani Vila et al, 1979; Ribier et al,
1984) (Fig. 1), and the delivery of two agents, methotrexate (MTX) and
doxorubicin (DOX). It has been possible to study vesicles of different
size, composition and properties, although both Surfactants I and II are
mixtures with n being an average value. Some progress towards optimization
of the systems has been made, bearing in mind that it is unlikely that any
one system will be ideal for all drugs and delivery objectives, because of
the effect of drug on the properties of the vesicle, and because of the
differences in the affinities of carrier and target organs. The more recent
choice of the monodisperse, pure analogue, Surfactant IV (Fig. 1) has led to
formulation difficulties due to the tendency of the vesicles to flocculate
on preparation.

The omission of cholesterol from non-ionic surfactant vesicles had
curiously little effect on the pharmacokinetics of DOX, in spite of the fact

Table 1. Some Vesicle-Forming Non-Ionic Surfactants

Surfactant	Reference
Alkyl polyglyceryl ethers	Vanlerberghe et al, 1978*
Alkyl Polyoxyethylene ethers	Usselmann and Muller-Goymann 1984* Chauhan and Lawrence, 1989 Hofland et al, 1989
Alkyl PEG polyglycerol	Muller-Goymann, 1987
Dialkyl polyglyceryl ethers	Vanlerberghe et al, 1972; 1978
Dialkyl polyoxyethylene ethers	Okahata et al, 1981
Steroidal oxyethylene ethers	Vanlerberghe et al, 1978* Patel et al, 1984
Steroidal lariat ethers	Echegoyan et al, 1988

*See also the following L'Oreal patents:
G. Kalopissis & G. Vanlerberghe (1967) French Patent No. 1, 477,048
G. Kalopissis & G. Vanlerberghe (1969) U.K. Patent No. 1, 155,712
G. Vanlerberghe & R.M. Handjani (1975) French Patent No. 2, 358,991
G. Vanlerberghe & H. Sebag (1979) French Patent No. 2, 465,78
G. Vanlerberghe & R.M. Handjani, 1979, U.K. Patent No. 1, 539,625
G. Vanlerberghe & R.M. Handjani, 1982, U.K. Patent No. 2, 013,609

that, without cholesterol, the vesicles are more permeable (Dr. R.M. Handjani-Vila, L'Oreal, Paris, maintains that preparations without cholesterol (or an equivalent molecule) cannot be considered to be vesicles as the surfactants will be present below the phase transition temperature; personal communication). The incorporation of cholesterol into the bilayer reduces the efflux of entrapped solute (Khand et al, 1987) and stabilizes the system against degradation (Rogerson et al, 1987). The temperature-dependency of drug release, which may be utilized in localized treatment by hyperthermia, is similar to that exhibited by liposomes, but both are strongly dependent on the particular lipid used and its phase transition temperature. Cholesterol abolishes the clear temperature dependency of flux; the incorporated drug may also affect the phase transition temperature. However, DOX levels in S180 tumours are higher following delivery in NSVs prepared with 50 mol % cholesterol, a result reflected in the slower growth of tumour (Rogerson et al, 1988; Kerr et al, 1988).

The structural integrity of the bilayer can be maintained yet the surface characteristics of the NSVs altered, by the incorporation of cholesteryl poly(n)oxyethylene ethers (n=5, 16 and 24) (Fig. 2) into the vesicles. The most hydrophilic additive (n=24) does not prevent uptake by the liver as had been hoped. However, not all NSVs are sequestered by the liver; the relationship between composition and destination have not been unravelled. Adsorption of DOX is clearly demonstrated by the modification of its distribution following mixing with empty NSVs. Adsorbed drug will, of course, influence the surface properties of the vesicle. While vesicles are versatile because of the ease of compositional change, the alteration of one component tends to affect several physical parameters. The addition of stearylamine to NSVs has resulted in binding of DOX in or to the vesicle so

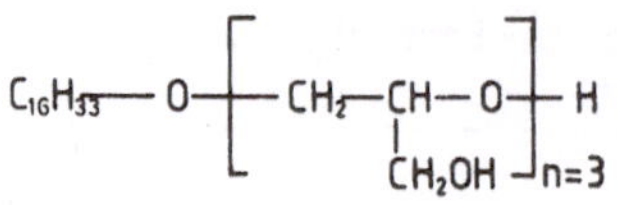

$C_{16}H_{33}-O-[-CH_2-CH-O-]_{n=3}-H$ with CH_2OH

Surfactant I, m.wt., 473

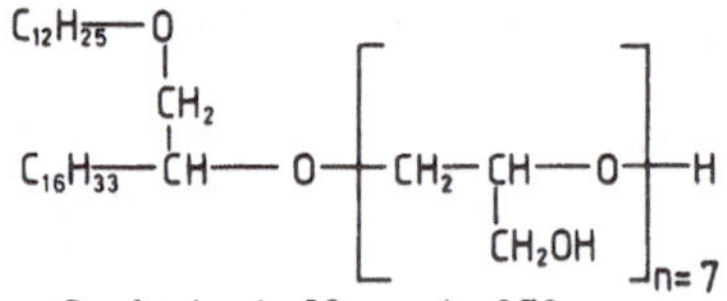

$C_{12}H_{25}-O$
CH_2
$C_{16}H_{33}-CH-O-[-CH_2-CH-O-]_{n=7}-H$ with CH_2OH

Surfactant II, m.wt., 972

$C_{16}H_{33}-O-[-CH_2-CH-CH_2-O-]_2-H$ with OH

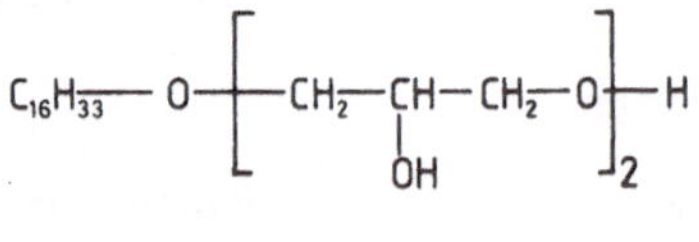

Surfactant IV, m.wt., 390

Fig. 1. Structure of the surfactants used in this work.

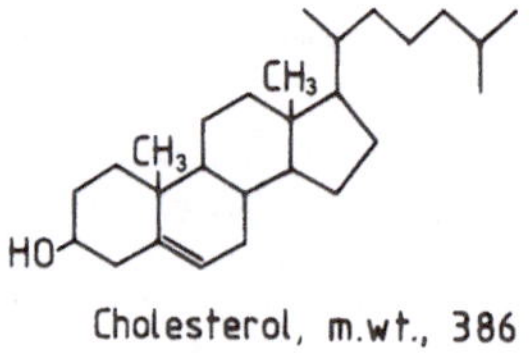

Cholesterol, m.wt., 386 7

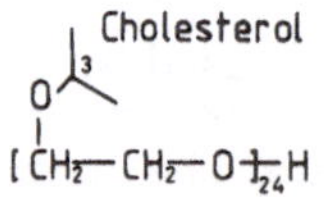

Cholesterol
O ·3
$[CH_2-CH_2-O-]_{24}H$

Average m.wt., 1443

HLB 14

Solulan C-24

Cholesterol $-O-[-CH_2-CH_2-O-]_{16}-H$

Estimated m.wt., 1091   HLB 15

Solulan 16

Cholesterol $-O-[-CH_2-CH_2-O-]_5-H$

Estimated m.wt., 607   HLB 8

Solulan 5

Fig. 2. Structure of the cholesteryl polyoxyethylene ethers used in this work.

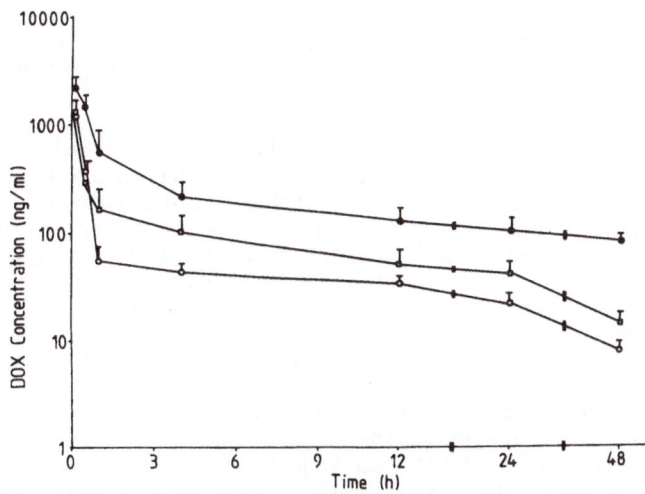

Fig. 3. Plasma concentrations of DOX as a function of time after bolus injection of 5 mg per kg into the tail vein of AKR mice of free drug in solution (open circles), empty vesicles given in conjunction with free drug (open squares) or DOX entrapped in vesicles (closed circles) composed of I 50:CHOL 25:SOL24 25. The vertical bar denotes the standard deviation of the drug concentration.

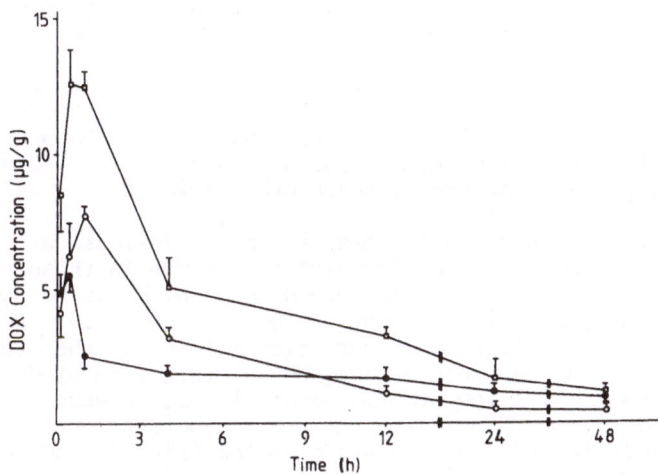

Fig. 4. Heart concentrations of DOX as a function of time after bolus injection of 5 mg per kg into the tail vein of AKR mice of free drug in solution (open circles), empty vesicles given in conjunction with free drug (open squares) or DOX entrapped in vesicles (closed circles) composed of I 50:CHOL 25:SOL24 25. The vertical bar denotes the standard deviation of the drug concentration.

that the system is virtually inactive. From such systems have emerged clearer indications of the effect of delivery rate on drug activity and on metabolism, which is affected by its mode of delivery (Rogerson et al, 1988).

In applications where uptake by the reticuloendothelial system is a disadvantage, the nature of the surface of colloidal spheres and liposomes is crucial. Converting hydrophobic microspheres into hydrophilic microspheres by adsorption of the non-ionic block copolymer surfactants, the poloxamers, serves to divert the carriers from the liver, phagocytic uptake of polystyrene decreasing with increasing thickness of the adsorbed chain (Ilum et al, 1987). One might therefore expect some differences in uptake of surfactant I and II vesicles, although the difference in hydrophilic chain length (n=3 and n=7, respectively) is relatively small.

There will be some obvious differences in the behaviour of non-ionic surfactant vesicles and phospholipid-based vesicles _in vivo_: the suggested transfer of lipid that occurs between injected liposomes and high-density lipoprotein, for example, is less likely to take place, although there may well be some interaction. Certain non-ionic surfactants intercalate with low-density lipoprotein particles _in vitro_ causing changes in diameter, possibly by inducing the unfolding of the apoprotein (Tucker and Florence, 1983) but the extent to which this (and indeed any biological activity due to the surfactant) is likely, is dependent on the freedom of the niosomal surfactant to be present as monomer. Vesicle-monomer equilibria are not well defined.

Pharmacokinetic studies in mice have demonstrated data now typical of vesicular delivery (Fig. 3) where plasma levels are maintained for longer periods after administration of the drug in vesicle form. The admixture of empty vesicles and drug prior to administration has demonstrated that there is sometimes weak and sometimes more extensive binding of DOX to the exterior of the vesicles. The data on mixed systems are of some interest, and it is clear that the simple admixture leads to changes in the distribution of DOX to other organs, eg. the heart (Fig. 4). The decreased cardiac levels of drug with both vesicular and mixed systems reflects an advantage of vesicular delivery noted with liposomes. If this can be reproduced in patients it would indicate an increase in therapeutic ratio.

In adding cholesteryl ethers to the vesicular structure to increase hydrophilicity, further complications of the influence of the cholesteryl ethers on size, charge and solute carrying capacity have been revealed (Cable et al, 1989a, b). The primary purpose to avoid the RES did not seem to be borne out, perhaps because of these other effects.

Because of difficulties in the preparation of vesicles with Surfactant IV, stearylamine was added to the formulation; it was in these studies that rather complex patterns of behaviour were noted, which can be partially attributed to binding and slow release of the drug (Fig. 5). Apparently a complex is formed which allows only very slow release of the drug. Plasma levels are virtually as we would expect, but distribution of drug into peripheral tissues is of a low order. Levels in most tissues following administration of DOX in NSVs is low, as are metabolite levels. Of course the levels in implanted ROS tumours are also low (Fig. 6) and tumour growth is barely affected by the treatment. However, this work demonstrates the ability to manipulate the composition of NSVs to control release rate over a wide range. The effect of composition on drug release is shown in Fig. 7.

During this work histological evidence (from fluorescence migrographs of implanted ROS tumours) of the deposition of vesicles in tumour vessels was obtained. While the evidence for extravasation is not by any means

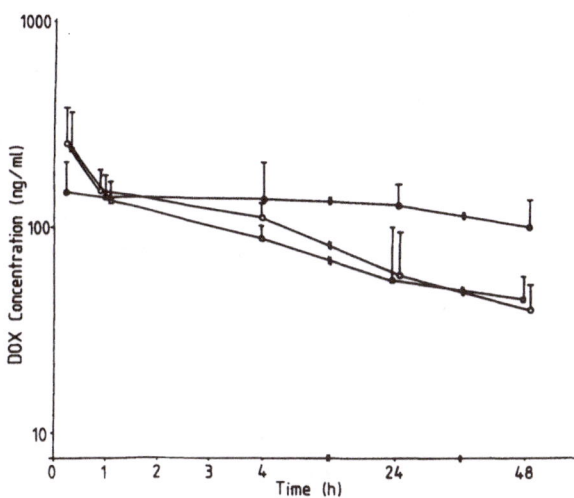

Fig. 5. Plasma concentrations of DOX as a function of time after bolus injection of 10 mg per kg into the tail vein of AKR mice of free drug in solution (open circles), empty vesicles given in conjunction with free drug (open squares) or DOX entrapped in vesicles (closed circles) composed of IV 47.5:CHOL 47.5:SA 5.0. The vertical bar denotes the standard deviation of the drug concentration.

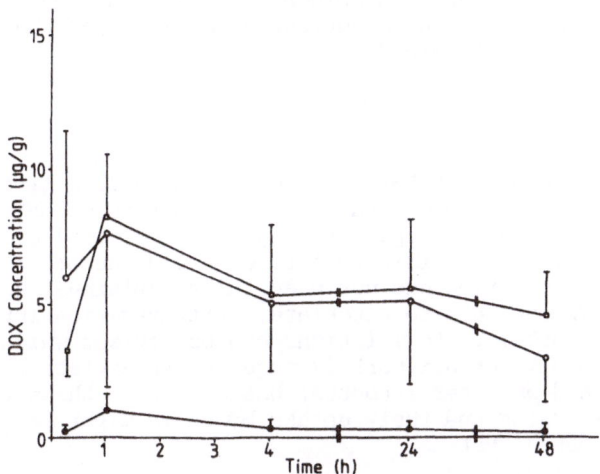

Fig. 6. ROS tumour concentrations of DOX as a function of time after bolus injection of 10 mg per kg into the tail vein of AKR mice of free drug in solution (open circles), empty vesicles given in conjunction with free drug (open squares) or DOX entrapped in vesicles (closed circles) composed of IV 47.5:CHOL 47.5:SA 5.0. The vertical bar denotes the standard deviation of the drug concentration.

clear, it is evident that the fluorescence several hours after administration suggests some affinity of the vesicles for the tumour vessels. At 22 h the fluorescence is more diffuse, suggesting the slow release from deposited vesicles. Presant et al (1990) have recently imaged Kapovi sarcoma with indium-labelled liposomes in AIDS patients and suggest proliferation of abnormal blood vessels in these tumours (Friedman-Kine et al, 1980)

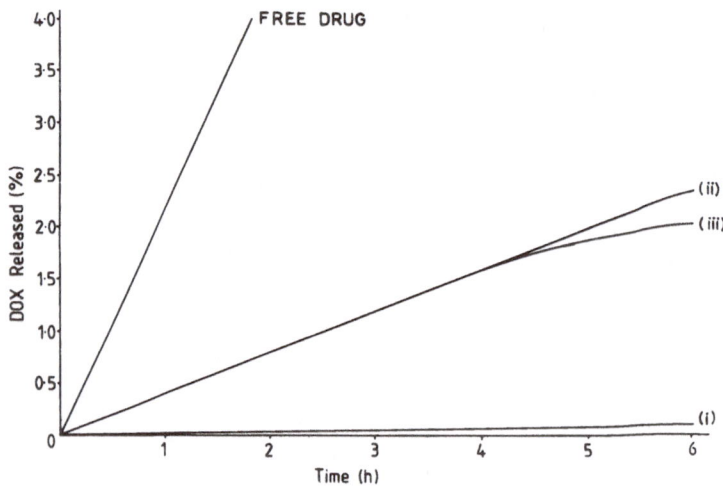

Fig. 7. Release of DOX across a dialysis membrane from a free drug solution
(1 mg/ml, 5 ml) and drug loaded vesicles composed of i) IV
47.5:CHOL 47.5:SA 5.0, ii) IV 49.5:CHOL 49.5:SA 1.0 and iii) IV
85.5:CHOL 9.5:SA 5.0.

and that extravasation does or can occur. The notion that all vesicles are
scavenged rapidly by the liver and spleen is perhaps not borne out by the
data from our work, although the quantitative estimate of the deposition of
surfactant has not been determined.

CONCLUSIONS

     We have so far found that NSVs behave very much like liposomes (if such
generic comparisons can be made); the non-ionic nature of NSVs is more
apparent than real, and some (e.g. Surfactant IV vesicles) are sensitive to
electrolyte addition. Their affinity for, or avoidance of, the liver
following i.v. administration has not been fully explained, but some studies
do suggest that NSVs continue to circulate. Extravasation into tumour
tissue has not been proved. Formulations can be devised which release drug
very slowly. Certainly further work is required to explore their true
potential by oral and parenteral routes: basic work on the affinity of the
vesicles for vessel walls and their uptake by cells might reveal differences
from phospholipid-based systems.

REFERENCES

Azmin, M.N., Florence, A.T., Handjani-Vila, R.M., Stuart, J.F.B.,
     Vanlerberghe, G. and Whittaker, J.A., 1985, The effect of non-ionic
     surfactant vesicle (biosome) entrapment on the absorption and
     distribution of methotrexate in mice, J.Pharm.Pharmacol., 37:237.
Azmin, M.N., Florence, A.T., Handjani-Vila, R.M., Stuart, J.F.B.,
     Vanlerberghe, G. and Whittaker, J.S., 1986, The effect of niosomes
     and polysorbate 80 on the metabolism and excretion of methotrexate in
     the mouse, J.Microencap., 3:95.
Baillie, A.J., 1988, Niosomes: a putative drug carrier system, in:
     "Targeting of Drugs", G. Gregoriadis and G. Poste, eds., Plenum, New
     York.
Baillie, A.J., Florence, A.T., Hume, L.R., Muirhead, G. and Rogerson, A.,

1985, The preparation and properties of niosomes - non-ionic surfactant vesicles, J.Pharm.Pharmacol., 37:863.

Baillie, A.J., Coombs, G.H., Dolan, T.F. and Laurie, J., 1986, Nonionic surfactant vesicles, niosomes, as a delivery system for the antileishmanial drug, sodium stibogluconate, J.Pharm.Pharmacol., 38:502.

Burkhanov, S.A., Kosykh Repin, V.S., Saatov, T.S., Torchilin, V.P., 1988, Interaction of liposomes of different phospholipid and ganglioside composition with rat hepatocytes, Int.J.Pharmaceutics, 46:31.

Cable, C. and Florence, A.T., 1988, Mixed poly(glycerol)-poly(oxyethylene) ether niosomes, J.Pharm.Pharmacol., 40:30P.

Cable, C., Cassidy, J., Kaye, S.B. and Florence, A.T., 1988, Doxorubicin in cholesteryl polyoxyethylene ether modified niosomes: evidence of enhancement of absorption in mice? J.Pharm.Pharmacol., 40:31P.

Carter, K.C., Baillie, A.J., Alexander, J. and Dolan, T.F., 1988, The therapeutic effect of sodium stibogluconate in BALB/c mice infected with Leishmania donovani is organ-dependent, J.Pharm.Pharmacol., 40:370.

Chauhan, S. and Lawrence, M.J., 1989, The preparation of polyoxyethylene containing non-ionic surfactant vesicles, J.Pharm.Pharmacol., 41:6P.

Echegoyan, L.E., Hernandez, J.C., Kaifer, A.E., Gokel, G.W. and Echegoyan, L., 1988, Aggregation of steroidal lariat ethers: the first example of nonionic liposomes (niosomes) formed from neutral crown ether compounds, J.Chem.Soc.Chem.Commun., 836.

Florence, A.T. and Baillie, A.J., 1989, Nonionic surfactant vesicles - alternatives to liposomes in drug delivery? in: "Novel Drug Delivery and its Therapeutic Applications", L.F. Prescott and W.S. Nimmo, eds., Wiley, Chichester.

Friedman-Kien, A.E., Laubenstein, L.J., Rubinstein, P., 1980, Disseminated Kaposi's sarcoma in homosexual men, Ann.Intern.Med., 96:693.

Handjani-Vila, R.M., Ribier, A., Rondot, B. and Vanlerberghe, G., 1979, Dispersion of lamellar phases of nonionic lipids in cosmetic products, Int.J.Cosmet.Sci., 1:303.

Hofland, J.E.J., Bowstra, J.A., Hodde, H., Spres, F., Junginger, H.E., 1989, Nonionic surfactant vesicles in transdermal formulations: controlled release and in vitro effects on human skin, Pharm.Res., 6:S.178.

Hunter, C.A., Dolan, T.F., Coombs, G.H. and Baillie, A.J., 1988, Vesicular systems (niosomes and liposomes) for delivery of sodium stiboglucamate in experimental murine visceral leishmaniasis, J.Pharm. Pharmacol., 40:161.

Illum, L., Jacobsen, L.O., Muller, R.H., Mak, E. and Davis, S.S., 1987, Surface characteristics and the interaction of colloidal particles with mouse peritoneal macrophages, Biomaterials, 8:113.

Kerr, D.J., Rogerson, A., Morrison, G.J., Florence, A.T. and Kaye, S.B., 1988, Antitumour activity and pharmacokinetics of niosome encapsulated adriamycin in monolayer spheroid and xenograft, Br.J.Cancer, 58:432.

Khand, L., Rogerson, A., Halbert, G.W., Baillie, A.J. and Florence, A.T., 1987, The effect of cholesterol on the release of doxorubicin from nonionic surfactant vesicles (niosomes), J.Pharm.Pharmacol., 39(Suppl.):41P.

Muller-Goymann, C.C., 1987, The influence of the microstructure on consistency and physical stability of an o/w cream, Acta Pharmaceutica Technologica, 33:60.

Okahata, Y., Tanamuchi, S., Nagai, M. and Kunitake, T., 1981, Synthetic bilayer membranes prepared from dialkyl amphiphiles with nonionic and zwitterionic head groups, J.Colloid Interface Sci., 82:401.

Patel, K.R., Li, M.P., Schuh, J.R. and Baldeschwieler, J.D., 1984, Pharmacological efficacy of a rigid non-phospholipid liposome drug delivery system, Biochim.Biophys.Acta, 797:20.

Presant, C.A., Blayney, D., Proffitt, R.T., 1990, Preliminary report: imaging of Kaposi sarcoma and lymphoma in AIDS with indium-111-labelled liposomes, Lancet, i:1303.

Ribier, A., Handjani-Vila, R.M., Bardez, E. and Valeur, B., 1984, Bilayer fluidity of non-ionic vesicles. An investigation by differential polarized phase fluorometry, Colloids and Surfaces, 10:155.

Rogerson, A., Cummings, J. and Florence, A.T., 1987, Adriamycin-loaded niosomes: drug entrapment, stability and release, J.Microencap., 4:321.

Rogerson, A., Cummings, J., Willmott, N. and Florence, A.T., 1988, The distribution of doxorubicin in mice following administration in niosomes, J.Pharm.Pharmacol., 40:337.

Stafford, S., Baillie, A.J. and Florence, A.T., 1988, Drug effects on the size of chemically defined non-ionic surfactant vesicles, J.Pharm. Pharmacol., 40:26P.

Tucker, I.G. and Florence, A.T., 1983, Interactions of ionic and nonionic surfactants with plasma low density lipoprotein, J.Pharm.Pharmacol., 35:705.

Usselmann, B. and Muller-Goymann, C.C., 1984, Struktureller aufbau von Cholesterol-Polyoxyathylenfettalkoholather-Wasser-Mischungen, Progress in Colloid and Polymer Sci., 69:56.

Vanlerberghe, G., Handjani-Vila, R.M., Berthelot, C. and Sebag, H., 1972, Synthese et activite de surface comparee d'une serie de nouveaux derives non-ioniques, Proceedings of the 6th International Congress on Surface Activity, Carl Hanser Verlag, Munich.

Vanlerberghe, G., Handjani-Vila, R.M. and Ribier, A., 1978, Les "niosomes" une nouvelle famille de vesicules a base d'amphiphiles non ioniques, Colloques Nationaux du C.N.R.S., No. 938.

TARGETED THERAPY OF LIVER CARCINOMA

N.I. Markham and J.R. Novell

Academic Department of Surgery, Royal Free Hospital
London, UK

INTRODUCTION

This chapter is concerned with the clinical aspects of anti-cancer drug targeting with respect to liver tumours.

Ever since Ehrlich proposed a "magic bullet" for the specific eradication of disease within the body, scientists have sought to realise his dream in the treatment of solid tumours. The concept is simple and elegant - that highly tumour-toxic treatment could be given which would "home-in" on tumour tissue to destroy it, whilst leaving other tissue unaffected.

The tumouricidal agent must, of course, be delivered by a suitable vehicle - itself non-toxic but capable of carrying with it a "warhead" which would be released at the target site. However, whilst the theory is straightforward, the practice is far from easy.

The most significant hurdle to overcome is that of finding a suitable delivery system. In its simplest form, site-specific dispatch of a drug can be achieved by merely infusing it into the vascular supply of the organ or tumour itself. This might be considered a form of "mechanical" targeting. In all but a few situations, however, this is impractical. Firstly, the vessels feeding the tumour are often not easily accessible and secondly active drug may escape into the general circulation after its first pass through the tumour.

The most potentially promising targeting system would make use of the concept of "tumour associated antigens". It has long been suspected that tumour cell surfaces possess antigens peculiar to the tumour tissue and therefore different to that of the native cell membrane (Old, 1981). Antibodies raised against these antigens and linked to suitable toxic warheads would recognise and then bind to their ligands to produce cell damage. Unbound antibody-warhead complexes would be eliminated or excreted without causing any cell damage. Once again, the theory outstrips the practice, since initially only impure mixtures of antibodies could be raised, resulting in crossreactivity with non-tumour tissue. The advent of monoclonal antibodies, however, raised great hopes that this problem would be considerably reduced (Kohler and Milstein, 1975; Mach et al, 1981).

Targeting of Drugs, Edited by G. Gregoriadis *et al.*
Plenum Press, New York, 1990

Other targeting systems have also been explored, and one in particular has shown considerable promise in the treatment of liver tumours. The fortuitous discovery that an oil used in conventional imaging of lymphatic channels is also selectively retained by hepatic tumour tissue has resulted in the development of a second unique targeting system (Nakakuma et al, 1979).

PRIMARY LIVER CANCER

Although comparatively rare in Europe, primary hepatocellular carcinoma (PLC) is one of the commonest and most malignant tumours in the world. There is evidence that the incidence is increasing in both Western Europe (Burnett et al, 1978) and in the USA (Peters et al, 1977).

Surgical resection remains the only hope of cure, but perhaps only 10% or less are suitable for surgery, either because the disease is too extensive within the liver, or has already metastasised. For those patients with an irresectable tumour the prognosis is bleak, the mean survival being three months (Sherlock, 1985). A wide variety of non-surgical treatments have been claimed to provide a measure of symptomatic palliation. These include devascularisation techniques, radiotherapy and chemotherapy (used individually or in combination), cryotherapy and alcohol injection (Wheeler et al, 1979; Ariel,1965; Johnson et al, 1978; Gilbert et al, 1985 and Livraghi et al, 1988). There is, however, little evidence that any significantly improves survival.

This chapter will briefly explore the possibilities of targeted treatment using two different guidance systems.

ANTIBODY-GUIDED TARGETING OF HEPATIC TUMOURS

If tumour cells do indeed possess antigens on their cell surface which are not expressed by their normal counterparts, then monoclonal antibody (MCA) technology provides a theoretical means of being able to target toxic therapy to the tumours. The antibodies, being monoclonal, are pure solutions of single antibodies and should not recognise targets other than the specific tumour antigens. They may be attached to drugs, radioisotopes or toxins or even used on their own as a means of active immunotherapy.

Initial attempts at producing MCAs specific for primary hepatocellular carcinoma were encouraging. Some of the gamma camera pictures obtained after intravenous administration of radiolabelled antibody showed excellent tumour localisation, including images of extra-hepatic spread of disease (Markham et al, 1986) (Fig. 1). However, it became clear that cross-reactivity reactions with antigens also expressed on the surface of cirrhotic cells were occurring and this lack of true tumour specificity presented a major obstacle to successful targeting (Markham et al, 1986). Paradoxically, it is work with polyclonal anti-ferritin antibodies that has so far produced the most impressive results. Order et al (1985) have treated 105 patients using radiolabelled anti-ferritin, a polyclonal antibody linked with I-131 as the cytotoxic warhead. Further work from their unit describes seven patients considered irresectable on the grounds of size, spread or location of tumour who were treated with the same antibody complex and whose tumours shrunk to a size where they could be successfully resected (Sitzmann et al, 1987).

Using 90-Yttrium as the cytotoxic warhead theoretically offers an advantage in delivering higher dose rates and greater total doses to the targeted tumour. Initial reports of Phase I studies suggest that this is

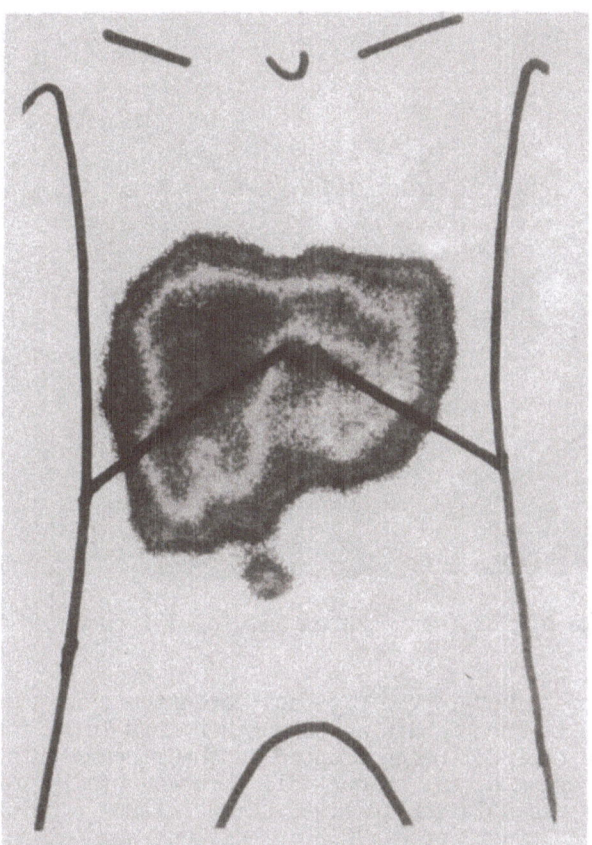

Fig. 1. Gamma scintigraphy of the liver following intravenous administration of I-131-labelled YPC2-/38.8 monoclonal in a patient with
PLC: tumour is demonstrated in both hepatic lobes and para-aortic
nodes.

possible (Order et al, 1988), although crossreactivity with the immunolog-
ically similar ferritin in non-neoplastic liver, heart and spleen may again
cause problems. The logical extension of this work is to use monoclonal
anti-ferritin and provisional reports have testified to better tumour dose
deposition in an animal model (Zhang et al, 1988). Future developments may
include the development of MCAs raised in humans to avoid the problems of
formation of human antibody to the MCAs, which are currently raised in
animals (Sikora et al, 1985).

CHEMICAL TECHNIQUES OF TARGETING LIVER TUMOURS

Lipiodol Ultra Fluid (May and Baker Pharmaceuticals, Dagenham, England)
is a stable iodine addition product of the ethyl esters of saturated and
unsaturated fatty acids (stearic, palmitic, oleic, linoleic and linolenic
acids) derived from poppyseed oil: it contains 475 mg of iodine per ml (38%
by weight) and has been used for many years as a lymphographic contrast
medium. In 1979 Lipiodol was infused into the hepatic artery of a laboratory dog and selective retention of the oil was seen in foci of hepatocellular carcinoma (Nakakuma et al, 1979). Lipiodol has been used in this
way for clinical localization studies and has been found to be particularly
effective in demonstrating the small "daughter" nodules often found in
association with the main tumour mass of PLC (Hayashi et al, 1987) (Fig. 2).

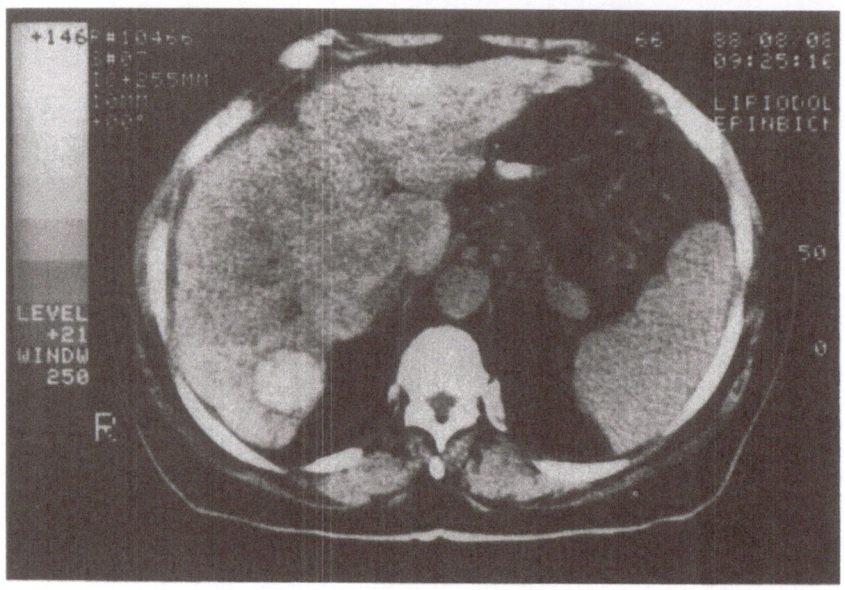

Fig. 2. Small nodule of PLC demonstrated on Lipiodol-enhanced CT scan.

The precise mechanisms of prolonged Lipiodol retention remain speculative. In particular, there is much debate as to whether this is an embolic phenomenon, with the Lipiodol being trapped in the microvasculature of the tumour circulation, or an actual tumour cell membrane binding or affinity. Patterns of Lipiodol accumulation in liver tumour tissue vary with respect to the tumour type (Kobayashi et al, 1987). Diffuse homogenous uptake is seen in PLC, which fits with the uniform hypervascularity of the tumour tissue. On the other hand, peripheral uptake with a central defect is usually the pattern of Lipiodol distribution seen in colorectal metastases in the liver, which commonly consist of a rim of dilated peritumoural sinusoids around a relatively avascular centre. These findings would tend to support the theory that Lipiodol retention is largely a function of microvascular entrapment. It has also been suggested that long term retention of Lipiodol may occur in the tumour extracellular space because of absent or poorly developed lymphatics (Iwai et al, 1984). Absorption of iodolipids from the tissues into the lymphatics is, however, very poor, and there is no evidence that the lymphatic system contributes significantly to the clearance of Lipiodol from normal hepatic tissue (Miller et al, 1987).

Lipiodol uptake by populations of tumour cells is also a possibility. Studies of the biodistribution of I-131 Lipiodol have shown marked activity in the spleen, bone marrow and lungs, compatible with uptake of the labelled lipid by the reticulendothelial system (Iwai et al, 1984). Also, indirect evidence of hepatocellular Lipiodol uptake is provided by experimental studies in which the major route of excretion from the liver was found to be via the bile (Iwai et al, 1987). Whatever the mechanism, however, it is clear that Lipiodol would appear to be an excellent vehicle for targeted therapy.

The first results using Lipiodol chemoembolisation were reported in 1983. Konno et al selectively infused the chemotherapeutic agent SMANCS (Styrene-maleic-anhydride-neocarzinostatin) mixed with Lipiodol in patients with PLC. They were able to show both a reduction in tumour size and of levels of circulating tumour marker in 13 out of 14 cases (Konno et al, 1983). A subsequent animal study demonstrated an increased survival

Table 1. Exclusion Criteria for Lipiodol Chemoembolisation

Age > 70 years

Serious concurrent medical illness

Previous chemotherapy

Bilirubin > 100 μmol/l

Haemoglobin < 10 g/dl

Severe clotting abnormalities

Hepatic failure

following infusion of the same combination when compared to Lipiodol alone (Iwai et al, 1984). A further prospective study of 124 patients with primary and secondary liver tumours demonstrated significantly greater survival following infusion of a Lipiodol and SMANCS combination than with hepatic artery ligation and chemotherapy (Tashiro and Maeda, 1985). Other authors have since reported good response rates and increased survival using Lipiodol in combination with various cytotoxic agents including doxorubicin and mitomycin C (Kobayashi et al, 1987; Ohishi et al, 1985).

Following an encouraging pilot study (Markham and Hobbs, 1988), we have now treated 25 patients with PLC at the Royal Free Hospital using a combination of Lipiodol and epirubicin administered at 2-3 monthly intervals. Epirubicin, a derivative of doxorubicin (4'-epi-doxorubicin), has been shown to have both an increased therapeutic index (Hochster et al, 1985) and increased activity against PLC (Epirubicin Study Group, 1987) compared to its parent drug.

All patients were fully assessed clinically, biochemically and radiologically, with special attention to liver function and the presence of extrahepatic spread. Patients were excluded if they met any of the criteria listed in Table 1.

In each patient the hepatic artery was selectively cannulated using the Seldinger technique, via the femoral artery. Ten ml of Lipiodol was emulsified with Epirubicin at a maximum dose of 90 mg/m^2, using the contrast medium Sodium Meglumine Ditrizoate (Urograffin, Schering Health Care, West Sussex, England) as an emulgent, and infused slowly under radiographic control. The patients were carefully monitored for signs of adverse reaction.

All tolerated the treatments well and the only observed side effects were mild and transient nausea, pyrexia and discomfort. Minimal but non-significant haematological depression and disturbance of enzymatic liver function was also seen, but these did not require treatment. One patient suffered temporary alopecia, but this was the only case where the expected side-effects of this dosage of chemotherapy, when given parenterally in the standard fashion, were seen. Computerized tomographic (CT) scans performed 10 days later showed retention of the Lipiodol within the liver tumour areas in all cases. This persisted for a further 2-3 months, when the treatment was repeated.

Analysis of tumour volumes as assessed by serial CT scanning suggest that tumour growth is being held in check, although significant reductions in size have not been seen (Figs. 3 and 4).

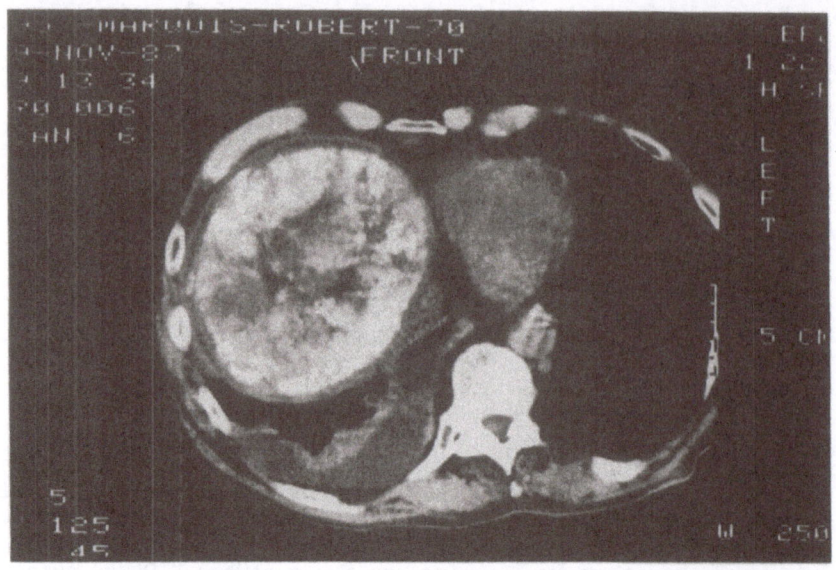

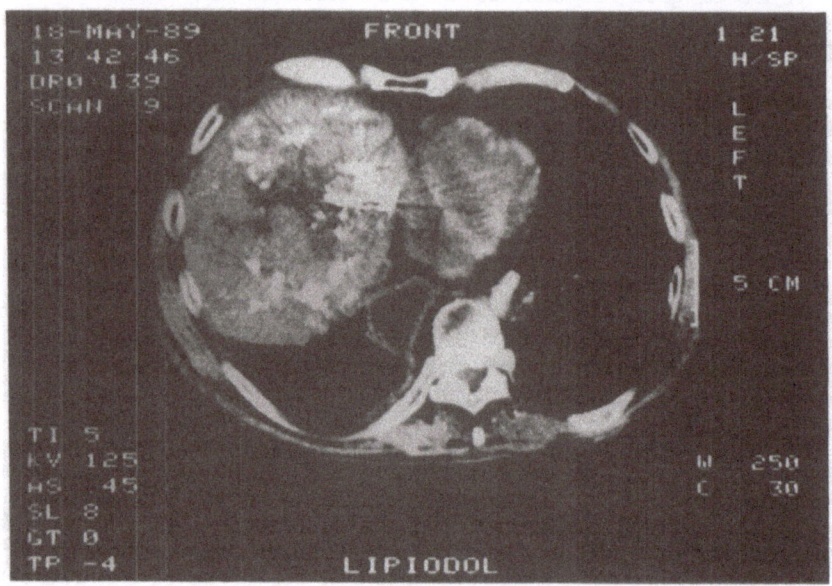

Figs. 3 and 4. CT scans of the liver taken 18 months apart, showing
Lipiodol deposition in a static focus of PLC.

Seventeen patients are currently evaluable: the mean survival is six
months from the start of treatment. Our longest survivor succumbed to an
unrelated condition after 21 months of treatment, with evidence of static
disease on serial CT scans and serum tumour markers.

The selective retention of Lipiodol by hepatic tumour tissue also makes
it an ideal vehicle for the delivery of radiotherapy. A fraction of the
iodine component of Lipiodol can be replaced in a simple exchange reaction
by I-131. Tumour uptake of the radiolabelled lipid has been studied using
scintigraphy, the I-131 Lipiodol exhibiting a similar preferential uptake
and retention to the non-radioactive lipid (Park et al, 1987; Raoul et al,

1988). In an animal tumour model it has been shown to have a significantly longer effective half life in tumour than in normal tissue and has resulted in almost complete necrosis of the tumour (Tsai et al, 1986). Early reports from clinical trials have also been encouraging, with reduction of both tumour size and ascites as well as symptomatic improvement in most cases. Tumour to non-tumour ratios of up to 20:1 have been achieved, and toxicity to surrounding liver parenchyma, bone marrow and lung has not been reported (Park et al, 1987; Bretagne et al, 1988).

We have now begun an evaluation of I-131 Lipiodol in PLC and have treated two patients with a mean administered activity of 540 MBq, infused as described above. As before, the treatments have been well tolerated and retention within tumour areas occurred with a tumour:non-tumour radioactivity ratio of 10:1 (unpublished data).

CONCLUSIONS

Primary liver cancer is one of the major causes of mortality throughout the world: only a small percentage of cases are amenable to surgery, and without it the prognosis is bleak.

Targeted delivery of cytotoxic therapy therefore offers an exciting hope for the future. Of the systems available, monoclonal antibodies have not as yet fulfilled their theoretical potential as "magic bullets" and as targeting vehicles they lack accurate specificity. However, the development of targeted chemotherapy using Lipiodol appears to offer great promise and it seems possible that targeted radiotherapy will extend the range of this technique. Larger clinical trials are needed to confirm these initial observations and these, as well as more detailed studies on the mechanism of Lipiodol binding to tumours are currently in progress.

REFERENCES

Ariel, I.M., 1965, Treatment of inoperable primary pancreatic and liver cancer by the intra-arterial administration of radioactive isotopes (Y-90 radiating microspheres), Ann.Surg., 162:267.
Bretagne, J., Raoul, J., Bourguet, P., Duvauferrier, R., Deugnier, Y., Faroux, R., Ramee, A., Herry, J-Y. and Gastard, J., 1988, Hepatic artery injection of I-131-labelled Lipiodol. Part II: Preliminary results of therapeutic use in patients with hepatocellular carcinoma and liver metastases, Radiology, 168:547.
Burnett, R.A., Patrick, R.S. and Spilg, W.C.S., 1978, Hepatocellular carcinoma and hepatic cirrhosis in the west of Scotland: a 25 year necropsy review, J.Clin.Pathol., 31:108
Epirubicin study group for hepatocellular carcinoma. Intra-arterial administration of epirubicin in the treatment of nonresectable hepatocellular carcinoma, Cancer Chemother.Pharmacol., 19:183, 1987.
Gilbert, J.C., Onik, G.M., Hoddick, W.K. and Rubinski, B., 1985, Real time ultrasonic monitoring of hepatic cryosurgery, Cryobiology. 22:319.
Hayashi, N., Yamamoto, K., Tamaki, N.,Shibata, T., Itoh, K., Fujisawa, I., Nakano, Y., Yamaoka, Y., Kobayashi, N., Mori, K., Ozawa, K. and Torizuka, K., 1987, Metastatic nodules of hepatocellular carcinoma: detection with angiography, CT, and US, Radiology, 165:61.
Hochster, H.S., Green, M.D., Speyer, J., Fazzini, E., Blum, R. and Muggia, F.M., 1985, 4'epidoxorubicin (Epirubicin): activity in hepatocellular carcinoma, J.Clin.Oncol., 3:1535.
Iwai, K., Maeda, H. and Konno, T., 1984, Use of oily contrast medium for selective drug targeting to tumour: enhanced therapeutic effect and X-ray image, Cancer Res., 44:2115.

Iwai, K., Maeda, H., Konno, T., Matsumara, Y., Yamashita, R., Yamasaki, K., Hirayama, S. and Miyauchi, Y., 1987, Tumour targeting by arterial administration of lipids: rabbit model with VX2 carcinoma in the liver, <u>Anticancer Res.</u>, 7:321.

Johnson, P.J., Williams, R., Thomas, H., Sherlock, S. and Murray-Lyon, I.M., 1978, Induction of remission in hepatocellular carcinoma with doxo-rubicin, <u>Lancet</u>, i:1007.

Kobayashi, H., Shimida, H., Yano, T., Maeda, T., Oyama, T. and Shinohara, S., 1987, Intra-arterial injection of adriamycin/mitomycin C Lipiodol suspension in liver metastases, <u>Acta Radiolog.</u>, 28:275.

Kohler, G. and Milstein, C., 1975, Continuous cultures of fused cells secreting antibody of predefined specificity, <u>Nature</u>, 256:495.

Konno, T., Maeda, H., Iwai, K., Tashiro, S., Maki, S., Morinaga, T., Mochinaga, M., Hiraoka, T. and Yokoyama, I., 1983, Effect of arterial administration of high-molecular-weight anticancer agent SMANCS with lipid lymphographic agent on hepatoma: a preliminary report, <u>Eur.J.Cancer Clin.Oncol.</u>,19:1053.

Livraghi, T., Salmi, A., Bolondi, L., Marin, G., Arienti, V., Monti, F. and Vettori, C., 1988, Small hepatocellular carcinoma: percutaneous alcohol injection - results in 23 patients, <u>Radiology</u>, 168:313.

Mach, J-P., Buchegger, F., Forni, M., Ritschard, J., Berchie, C., Lumbroso, J-D., Schreyer, M., Giradet, C., Accolla, R.S. and Carrel, S., 1981, Use of radiolabelled monoclonal anti-CEA antibodies for the detection of human carcinomas by external photoscanning and tomoscintiscanning, <u>Immunol.Today</u>, 2:239.

Markham, N., Ritson, A., James, O., Curtin, N., Bassendine, M. and Sikora, K., 1986, Primary hepatocellular carcinoma localised by a radio-labelled monoclonal antibody, <u>J.Hepatol.</u>, 2:25.

Markham, N.I. and Hobbs, K.E.F., 1988, Selective destruction of malignant hepatic tumours - the magic bullet approaches? <u>J.Roy.Coll.Surg.E.</u>, 33:205.

Miller, D.L., O'Leary, T.J. and Girton, M., 1987, Distribution of iodized oil within the liver after hepatic arterial injection, <u>Radiology</u>, 162:849.

Nakakuma, K. Tashiro, S., Uemura, K., Konno, T., Tanaka, M. and Yokoyama, I., 1979, Studies on the anticancer treatment with oily anticancer drug injected into the ligated hepatic artery for liver cancer. (Preliminary report) (in Japanese), <u>Nichidoku Iho</u>, 24(4):675.

Ohishi, H., Uchida, H., Yoshimura, H., Ohue, S., Ueda, J., Katsuragi, M., Matsuo, N. and Hosogi, Y., 1985, Hepatocellular carcinoma detected by iodized oil; use of anticancer agents, <u>Radiology</u>, 154:25.

Old, L.J., 1981, Cancer immunology: the search for specificity, <u>Cancer Res.</u>, 41:361.

Order, S.E., Stillwagon, G.B., Klein, J.L., Leichner, P.K., Siegelman, S.S., Fishman, E.K., Ettinger, D.S., Haulk, T., Kopher, K. and Finney, K., 1985, Iodine 131 antiferritin, a new treatment modality in hepatoma: a radiation therapy oncology group study, <u>J.Clin.Oncol.</u>, 3:1573.

Order, S.E., Vriesendorp. H.M., Klein, J.L. and Leichner, P.K., 1988, A phase I study of Yttrium-90 antiferritin: dose escalation and tumour dose, <u>Antibody, Immunoconj.& Radiopharm.</u>, 1:163.

Park, C.H., Suh, J.H., Yoo, H.S., Lee, J.T., Kim, D.I. and Kim, B.S., 1987, Treatment of hepatocellular carcinoma (HCC) with radiolabelled Lipiodol: a preliminary report, <u>Nucl.Med.Commun.</u>, 8:1075.

Peters, R.L., Afroudakis, A.P. and Tatler, D., 1977, The changing incidence of association of hepatitis B with hepatocellular carcinoma in California, <u>Am.J.Clin.Pathol.</u>, 68:1.

Raoul, J., Bourguet, P., Bretagne, J., Duvauferrier, R., Coornaert, S., Darnault, P., Ramee, A., Herry, J-Y. and Gastard, J., 1988, Hepatic artery injection of I-131-labelled Lipiodol. Part I: Biodistribution study results in patients with hepatocellular carcinoma and liver metastases, <u>Radiology</u>, 168:541.

Sherlock, S., 1985, "Diseases of the liver and biliary system", 7th edition, Oxford:Blackwell.

Sikora, K., Alderson, T., Nethersell, A. and Smedley, H., 1985, Tumour localisation by human monoclonal antibodies, Med.Oncol.& Tumor Pharmacother., 2:77.

Sitzmann, J.V., Order, S.E., Klein, J.L., Leichner, P.K., Fishman, E.K. and Smith, G.W., 1987, Conversion by new treatment modalities of non-resectable to resectable hepatocellular cancer, J.Clin.Oncol., 5:1566.

Tashiro, S and Maeda, H., 1985, Clinical evaluation of arterial adminis-tration of SMANCS in oily contrast medium for liver cancer, Jap.J.Med., 24:79.

Tsai, C., Kusumoto, Y., Harada, R., Shima, M., Nakata, K., Konno, K., Sato, A., Ishii, N., Koji, T. and Nagataki, S., 1986, Effect of intra-hepatic arterial infusion of I-131-labelled Lipiodol on hepato-cellular carcinoma of rat, Ann.Acad.Med.Singapore, 15:521.

Wheeler, P.G., Melia, W., Dubbins, P., Jones, B., Nunnerley, H., Johnson, P. and Williams, R., 1979, Non-operative arterial embolisation in primary liver tumours, Br.Med.J., 2:242.

Zhang, X.F., Klein, J.L. and Order, S.E., 1988, Quantitative comparison of tumour dose of radiolabelled monoclonal and polyclonal antibodies in an experimental tumour model, Antibody, Immunoconj.& Radiopharm., 1:35.

DRUG DELIVERY SYSTEMS: THE INDUSTRIAL VIEW

K. Yokoyama, Y. Ueda, A. Kikukawa and K. Yamanouchi

Research Division, Green Cross Corporation, Shodai-ohtani
2-1180-1, Hirakata, Japan 572

INTRODUCTION

One of the striking features of lipid emulsion particles is chylo-micron-like behavior in the circulation (Davis et al, 1985), i.e., high affinity to capillary walls, especially when the latter are inflamed (Shaw et al, 1979). This had led us to consider drug targeting against the walls of blood vessels and hence to pursue the development of lipid microsphere (fat emulsion) incorporating active substances. We have also become interested in adriamycin-macromolecule conjugates that are expected to increase the selectivity for tumor cells.

We wish to describe herein case histories dealing with microsphere formulations for dexamethasone (Yokoyama et al, 1985) and prostaglandin E_1 (Otomo et al, 1985) which we have recently launched on the Japanese market, together with the results of our recent studies on an adriamycin-oxidized dextran conjugate and its monoclonal antibody derivative.

INTRAVENOUS ADMINISTRATION OF DRUG DELIVERY SYSTEMS (DDS)

The relationship between the intended effects of a drug and its toxi-city is always of concern. The most important consideration in developing drugs is thus to find compounds which possess low toxicity as well as potent pharmacological activity. In addition, our great concern is also for a choice of an optimum formulation for the drug to fit both specificity and degree of disease. One of the approaches to minimize toxicity and to en-hance activity at the same time would be to devise a DDS wherein a drug accumulates at a specific disease site.

As for the administration route, DDS may be roughly classified into oral, topical and parenteral formulations as discussed elsewhere (e.g. Davis et al, 1985). The drug administered by any route will finally encounter the reticulendothelial system (RES) in most cases after entering the blood stream. With this in mind, the parenteral route, particularly the intra-venous (i.v.) and intra-arterial (i.a.), would be favorable to oral and topical administration, because this route would possibly avoid biodegrad-ation during the absorption processes. Subcutaneous and intramuscular administration, which is used typically for vaccines that need to be ab-sorbed rather gradually, would be of limited value if rapid onset of the

Table 1. Industrialization of DDS-Based Drugs in Injection Form

---

Economic points:

        Sold at high price
        Manufactured at low cost
        Patented

Technological points:

        Feasible for industrial scale production
        Reliable in shelf-life stability
        Reliable in sterilization

---

drug is needed. Intravenous administration leads to rapid onset of drug activity compared to i.a. administration, indicating suitability for enhanced targeting of the drug. We can thus conclude that i.v. administration would be the choice for DDS use.

DRUGS AND CARRIERS FOR DDS

Development of drugs by pharmaceutical industries must be time and cost effective (Table 1). Products to be marketed are usually expected to be manufactured at a low cost without patent difficulties and to be sold at a high sale price. In this regard, we have first chosen dexamethasone, prostaglandin E_1 and adriamycin which have been used in large quantities in a wide range of applications, but are limited in their uses because of serious side effects and instability in vivo.

Choice of the carrier is as important as the choice of the drug to be carried. Our top priority in choosing the carrier was to consider the feasibility for industrial scale production and safety. For technological reasons, it is desirable that the manufacturing technology of the carrier has been already established and a long shelf life guaranteed. Having considered these requirements, we first chose a fat emulsion. Liposomes would rank second in our hands. The use of fat emulsions and liposomes as carriers may be rationalized on the assumption that these particles would be localized in inflammatory lesions where macrophages are abundant and/or capillary permeability is increased (Tomlinson, 1986).

Fat emulsions have been manufactured on an industrial scale and used for nearly half a century and their safety has been already established (Davis et al, 1985). Liposomes are applicable to both water and lipid soluble drugs, whereas fat emulsions are only for lipid soluble drugs. Although recent progress in liposome technology is remarkable, there appears to be some room for improvement for large scale production, especially with reference to sterile filtration processes.

Unfortunately, fat emulsions are not suitable for adriamycin, and liposomal forms did not show satisfactory drug efficacy in our hands. Thus, conjugation with a macromolecule has been investigated. The macromolecule dextran was chosen because of its established reliability in clinical use.

EXAMPLES OF LIPID MICROSPHERE CONTAINING DRUGS

Dexamethasone Palmitate Incorporated in Lipid Microspheres

Dexamethasone belongs to a family of steroids with strong activity and has been used in large quantities in a wide range of clinical applications because of its anti-inflammatory and immunosuppressive effects. It is also known as a drug difficult to use because of serious side effects. Faced with the inherent drawbacks of the drug, we chose this steroid as the first targeting drug to be developed.

As discussed before, we favored fat emulsions as a carrier rather than liposomes. Thus the drug needed to be dissolved in the oil phase. The solubility of dexamethasone in soybean oil is not more than 0.01%. We found that a C_{16} alkyl ester (palmitoyl ester) of the drug is best suited for this purpose in terms of both solubility and effectiveness.

Manufacturing method. After dissolving dexamethasone palmitate into soybean oil, the oil phase was dispersed into the water phase with egg yolk phospholipids (YPL) by a Manton Gaulin homogenizer under a high pressure (300 kg/cm^2) at approximately 65°C. To the resultant emulsion was added glycerol, the mixture filtered through a 0.8 micron millipore filter, placed into 1 ml glass ampules and sterilized in an autoclave at 121°C. The final product has a pH of 7.5, with no particles larger than 1 micron. The composition and physicochemical properties are shown in Table 2.

Pharmacological studies. Figure 1 shows comparative data on the distribution of the steroid in adjuvant arthritis lesions in rats after i.v. injection of dexamethasone palmitate incorporated into lipid microspheres (designated hereafter as liposteroid) and of dexamethasone phosphate (DP) in an aqueous solution. With the exception of the first 30 minutes, we observed a nearly two-fold increase in tissue concentration with liposteroid throughout the experimental period (Yokoyama et al, 1985). Another study of the tissue distribution in animals (Mizushima et al, 1982) also demonstrated that the plasma level of liposteroid was higher than that of DP a few hours after i.v. administration, and that DP distributed more rapidly than did liposteroid into water-rich tissues such as muscle.

These results indicate a beneficial property of liposteroid for drug targeting as evidenced by the accumulation into inflamed tissues. A typical example obtained with carrageenin granuloma in rats (Yokoyama et al, 1985) is shown in Fig. 2. Liposteroid markedly suppressed the growth of granuloma, demonstrating 5.3 times greater anti-inflammatory activity than DP. Another study using adjuvant arthritis in rats (Yokoyama et al, 1985) also demonstrated that liposteroid was 3.3 times more effective than DP, suggesting that rheumatoid arthritis (RA) might be one of the candidate diseases for liposteroid treatment.

In order to clarify the mechanism of the high potency of liposteroid, in vitro, studies using rat macrophages were also carried out (Yokoyama et al, 1985). The rate of uptake of the steroid by rat peritoneal macrophages was significantly greater with liposteroid than with DP. Furthermore, the inhibitory effects of liposteroid on Fc receptor-dependent phagocytosis activity, superoxide anion release and chemotaxis by macrophages were more potent than those of DP. Therefore, the strong anti-inflammatory activity of liposteroid may be due not only to high accumulation into the inflamed lesions but also to the strong effects on macrophage functions. In addition, dexamethasone palmitate in liposteroid was found to exhibit its biological activity after forming free dexamethasone, which is initially produced by hydrolysis involving macrophages and monocytes (Okamoto et al, 1985; Ii et al, 1988).

Table 2. Composition of Liposteroid

Dexamethasone palmitate (250 mg as dexamethasone)	400 mg
Soybean oil	100 mg
Egg yolk phospholipids	12 mg
Glycerol	22.1 mg
Water for injection	q.s.
Total	1 ml

pH: 7.5; osmolarity: 280–300 mOsm: particle size: 200–400 nm average diameter; shelf life: two years at room temperature.

These studies suggest that liposteroid is taken up by the macrophages to a greater extent than free dexamethasone and strongly suppresses macrophage functions, thus increasing the effectiveness of the drug.

<u>Clinical studies</u>. Results of clinical studies conducted in Japan have been reported elsewhere, for instance a Phase I study (Miura et al, 1982), one related to vasoconstrictor activity (Sugai et al, 1984), therapeutic studies with RA patients (Kaneko, 1985) and a pharmacokinetic study with chronic RA patients (Ii et al, 1988).

Hoshi et al (1985) have reported the results of a multi-center double blind comparative trial that was conducted in 138 RA patients treated i.v. with liposteroid biweekly using DP in an i.m. dosage form (Decadron) as a reference drug and a fat emulsion in an i.v. dosage form (Intralipos) as placebo. Suppression of cortisol levels, which is considered a measure of drug side effects, was hardly observed when 1 to 3 ml of liposteroid (2.5–

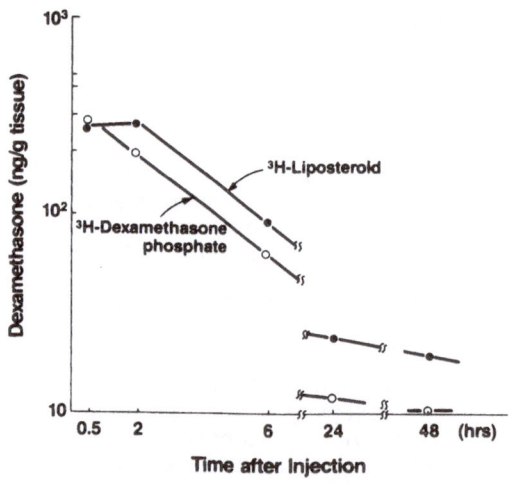

Fig. 1. Distribution of ^{3}H-dexamethasone palmitate given i.v. as lipid microspheres and ^{3}H-dexamethasone phosphate given i.v. as a solution to rats with arthritic lesion. Each drug was injected at a dose of 0.5 mg/kg as dexamethasone (25 µCi/kg) to arthritic female S.D. rats on day 13 after i.d. injection of 0.05 ml of an adjuvant mixture containing <u>Mycobacterium butyricum</u> into the volar surface of the right hind paw. Radioactivity was determined in the arthritic lesion of the left non-injected hind paw. (Taken from Yokoyama et al, 1985, with the permission of the copyright owner.)

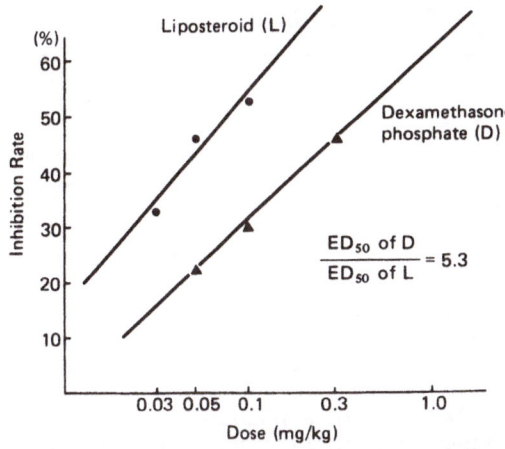

Fig. 2. Effects of liposteroid on carrageenin granuloma in rats. Rats with
carrageenin granuloma were given either liposteroid or dexamethasone
phosphate on day 5, 6 and 7 after injection of carrageenin. One
day after final injection of the steroid, the weight of granuloma
was measured. (Taken from Yokoyama et al, 1985, with the permission of the copyright owner.)

7.5 mg as dexamethasone) were injected biweekly into RA patients (Kaneko,
1985). Liposteroid (Limethason®) was accordingly approved on January 20,
1988 for rheumatoid arthritis.

Prostaglandin E_1 Incorporated in Lipid Microspheres

Prostaglandin E_1 (PGE1) has been used for peripheral vascular diseases
and diabetic neuropathy because of its strong vasodilating and anti-platelet
aggregation activities (e.g., Martin, 1980). However, PGE1 is also known to
have some disadvantages which make the drug difficult to use: i.e., rapid
inactivation of the drug in the lung because of an inherent property as an
autacoid hormone, and irritation of the vascular walls at the injection site
because of overly strong biological activities. In order to solve these
problems, the authors have pursued the development of an improved preparation of PGE1; that is, lipid microspheres containing PGE1 (designated hereafter as Lipo PGE1).

Preparation of lipo PGE1. The solubility of PGE1 in soybean oil is approximately 6 mg/100 g at 25°C. Although we prepared a wide range of ester
derivatives in an attempt to increase solubility, we could not find an ester
candidate which effectively balances the rate of hydrolysis with the rate of
onset of the drug's effects.

After dissolving PGE1 into soybean oil, the oil phase was emulsified
according to the method described for liposteroid, except that 1.8% (w/v)
YPL was used and 0.24% (w/v) oleic acid added as a stabilizer for autoclav-
ing. The composition and physicochemical properties are shown in Table 3.

Results of pharmacological studies. Most of our studies were conducted on
the basis of comparison with such established preparations as PGE1 cyclo-
dextrin (abbreviated hereafter as PGE1 CD) (e.g. Hamano et al, 1986; Nakura
et al, 1986; Esumi et al, 1986).

In Figs. 3 and 4, the hypotensive effects of lipo PGE1 are compared
with PGE1 CD using diabetic and spontaneously hypertensive rats (SHRs)
(Hamano et al, 1986). Responsiveness to vasoactive substances decreased

Table 3. Composition of Lipo PGE$_1$

Prostaglandin E$_1$	5 µg
Soybean oil	100 mg
Egg yolk phospholipids	18 mg
Oleic acid	2.4 mg
Glycerol	22.1 mg
Water for injection	q.s.
Total volume	1 ml

pH: 4.5–6.0; osmolarity: 280–300 mOsm; particle size: 200–400 nm average diameter; shelf life: 12 months at 4°C.

with the progress of diabetes. This phenomenon was observed with both PGE1 CD and isoproterenol, whereas with lipo PGE1, responsiveness changed with the progression of diabetes. The most striking difference between lipo PGE1 and PGE1 CD was seen in 10-week diabetic rats, where the depressor potency of lipo PGE1 was about 25 times as high as that of PGE1 CD. In the SHR model, lipo PGE1 was shown to be much more potent than PGE1 CD.

The high effectiveness (vasodilating activity) of lipo PGE1 is also supported by the fact that lipo PGE1 accumulates on the vascular walls (Nakura et al, 1986). Figure 5 shows comparative data on the vascular distribution of PGE1 given as lipo PGE1 and in an aqueous solution, in SHR and normal rats 5 min after i.v. injection. In normal rats, no significant difference between the two preparations was found. In the SHR model, however, lipo PGE1 accumulated on the vascular wall at a significantly higher level than free PGE1. These results suggest that the vascular walls of the SHR lesion may be one of the targeting sites of the lipid particles.

Clinical studies. Table 4 shows a summary of clinical studies which were conducted in Japan (Mizushima et al, 1983; Otomo et al, 1985). In the case of arterial duct-dependent congenital heart disease, lipo PGE1 showed a satisfactory effectiveness with a low incidence of adverse reactions. Although this study was not designed as a double blind one, the efficacy of lipo PGE1 was superior to that of PGE1 CD (Momma et al, 1984).

As for the other indications, all of which are related to the peripheral vascular disturbance, lipo PGE1 was significantly more effective than either PGE1 CD or inositol hexanicotinate (e.g. Mizushima et al, 1983). Accordingly, lipo PGE1 was approved by the Japanese authorities for the five conditions (Table 4) on June 28, 1988.

EXAMPLES OF ADRIAMYCIN-MACROMOLECULE CONJUGATES

Adriamycin-Oxidized Dextran Conjugate

Since antitumor agents usually have strong cytotoxicity against normal tissues as well as tumor tissues, amounts of antitumor agents sufficient to destroy tumor tissues can not be administered to cancer patients. However, therapeutic indices of antitumor drugs might be greatly improved if we can deliver the drugs to the tumor tissues more specifically. To this end, several modifications of antitumor agents have been carried out to improve their selectivity (Tomlinson, 1986). One typical example is to use the natural characteristics of drugs and tumors to induce accumulation of the drug into the tumor tissues. This is based on the finding (Iwai et al, 1984) that macromolecules tend to accumulate into the tumor tissues when introduced into the body because of the high permeability of the blood

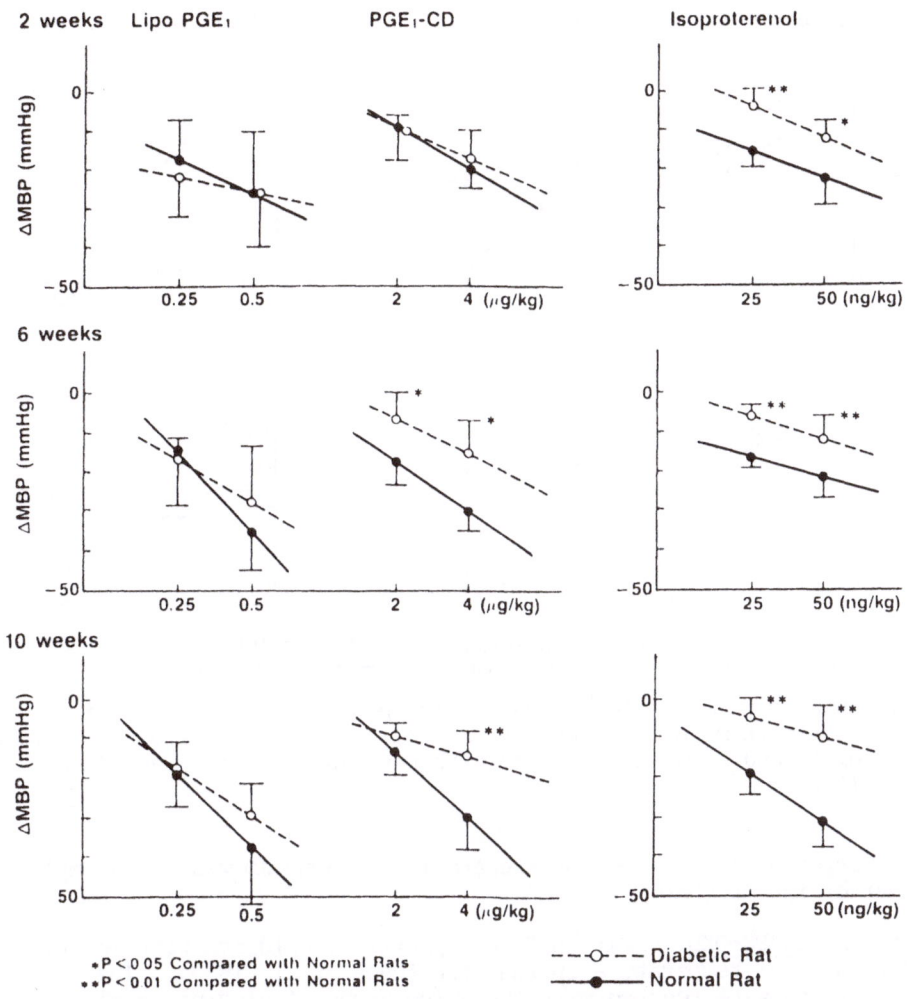

Fig. 3. Changes in mean blood pressure in diabetic rats. Lipo PGE$_1$ or reference drugs (PGE$_1$ CD and isopreterenol) were injected i.v. into rats 2, 6 and 10 weeks after giving streptozocin. Blood pressure was measured by a transducer under anesthesia with a mixture of urethane and α-choral. (Taken with permission from Hamano et al, 1986.)

vessels and the inadequate development of the lymph system in the tumor tissues.

Several years ago we attempted to make a normal IgG-adriamycin (ADM) conjugate based on the fact that normal IgG tends to be localized in the tumor tissues. In order to increase the ADM content in the conjugate, we used oxidized dextran (OXD) as a spacer. This IgG-ADM product showed stronger antitumor activity against rat experimental tumors than ADM and was also less toxic. We found that the plasma level of IgG-ADM-OXD was much higher than that of free ADM. Interestingly, the plasma level of the ADM-OXD which is an intermediate for IgG-ADM-OXD was found to be as high as that of IgG-ADM-OXD. We then turned our attention to the antitumor effects of ADM-OXD and found the same strong antitumor activity with ADM-OXD as with IgG-ADM-OXD. In addition, ADM-OXD was less toxic than ADM. From these results we decided to choose the ADM-OXD as a potent derivative for the

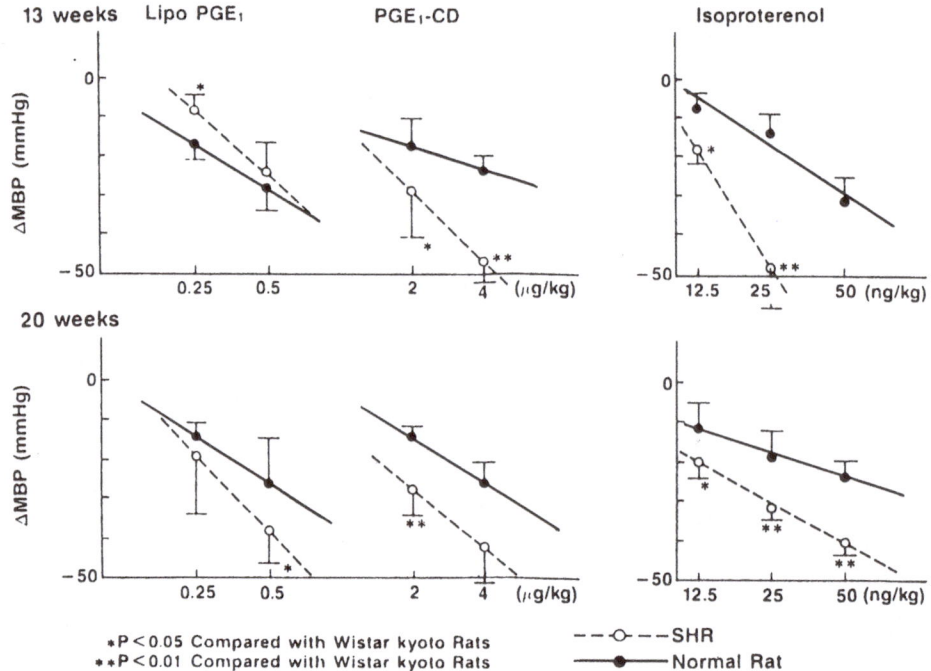

Fig. 4. Changes in mean blood pressure in spontaneously hypertensive rats. Rats with more than 160 mmHg systolic pressure were used at the age of 13 and 20 weeks. (Taken with permission from Hamano et al, 1986.)

passive targeting of ADM because preparation of ADM-OXD was much simpler than that of IgG-ADM-OXD.

Preparation of ADM-OXD. Dextran 70 was partially oxidized with sodium periodate at room temperature in the dark (Ueda et al, 1989). The resulting aldehyde groups were reacted with the amino group of glycine, yielding Schiff's base. After reducing the Schiff's base with sodium borohydride to stabilize the bond, the remaining glucose residues of the dextran-glycine were reoxidized with sodium periodate. The amino group of ADM was then reacted with the newly-formed aldehyde groups of the dextran-glycine to give ADM-OXD conjugate via the Schiff's base. The overall yield of ADM-OXD ranged from 70 to 80% based on ADM.

Characteristics of ADM-OXD. In contrast to free ADM, ADM-OXD was freely soluble in water due to the glycine moiety of the conjugate. The molecular weight of ADM-OXD was estimated to be 70 kd by gel permeation chromatography. The total amount of ADM in ADM-OXD was about 30% (w/w). ADM-OXD contained a small amount of free ADM (less than 2% (w/v) of the total ADM). As shown in Fig. 6, release of ADM from the conjugate increased under acidic conditions due to the characteristics of Schiff's base between ADM and the OXD backbone chain. The concentration of ADM-OXD also affected the release of ADM from the conjugate. Low concentration of ADM-OXD accelerated the release of free ADM. These characteristics are considered to be a great advantage of the conjugate if endocytosis is involved in its incorporation into tumor cells.

In vitro biological activity. Figure 7 shows the in vitro antitumor effect of ADM-OXD against MKN-45 human gastric adenocarcinoma. The effects of ADM-OXD were less pronounced than those of free ADM as long as the tumors were

Table 4.   Summary of Clinical Results on Lipo PGE₁

1. Arterial-duct-dependent conjenital heart disease

	No. of Patients	Efficacy	Dose	Adverse reaction
Lipo PGE₁	83	94.0%	5ng/kg/min.	30.1%
PGE₁ CD	(historical)	50 ~ 70%	50 ~ 100ng/kg/min.	70 ~ 80%

2. Buerger's diseases, arteriosclerosis obliterans (DBT)

	No. of Patients	Improvement*	Dose	Adverse reaction**
Lipo PGE₁	62	59.7%	10μg/day/4 weeks i.v.	9.4%
IHN	62	40.3%	1200mg/day/4 weeks p.o.	6.3%

IHN: Inositol Hexa Nicotinate   *p < 0.05, **N.S.

3. Diabetes-associated peripheral vascular and nervous disease

	No. of Patients	Improvement*	Dose	Adverse reaction**
Lipo PGE₁	84	60.7%	10μg/day/4 weeks i.v.	12.9%
PGE₁ CD	89	38.2%	40μg/day/4 weeks i.v.	23.8%

*p < 0.01, **N.S.

4. Collagenoses-associated peripheral vascular disturbances (DBT)

	No. of Patients	Improvement*	Dose	Adverse reaction**
Lipo PGE₁	66	54.5%	10μg/day/4 weeks	19.1%
Placebo	65	29.2%		16.4%

*p < 0.001, **N.S.

5. Vibration disease (Peripheral vascular & nervous disease) (DBT)

	No. of Patients	Improvement*	Dose	Adverse reaction**
Lipo PGE₁	49	85.7%	10μg/day/4 weeks i.v.	36.8%
PGE₁ CD	48	64.6%	40μg/day/4 weeks i.v.	58.2%

*p < 0.05. **N.S.   *p < 0.05, **p < 0.05

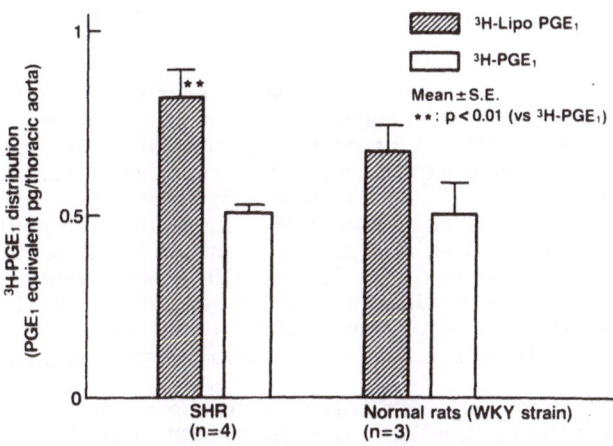

Fig. 5.   Distribution of ³H-PGE₁ in thoracic aorta of rats when injected with the lipid microsphere form of the drug or its aqueous solution form.   Spontaneously hypertensive 20-week old rats were given the drug at a dose of 2 μCi/10 ng as PGE₁/0.1 ml/kg/min and sacrificed after a 2 min infusion for measuring radioactivity. (Taken with permission from Nakura et al, 1986.)

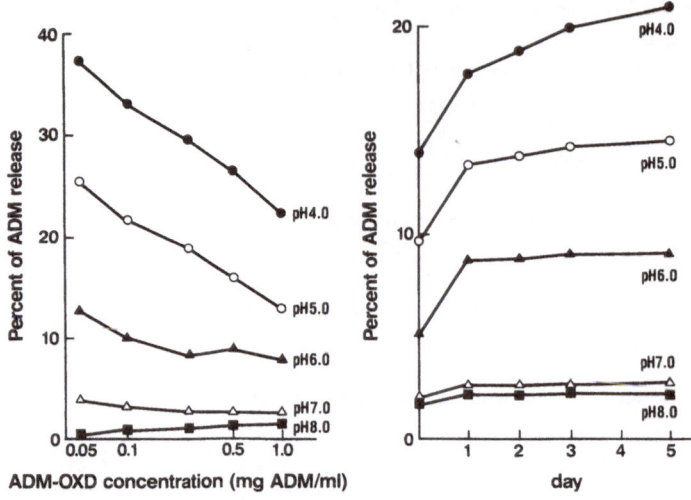

Fig. 6. pH stability of ADM-OXD.
         Left: ADM-OXD was dissolved in distilled water at various
         concentrations and then mixed with equal volume of McIlvain's buffer
         at a pH range of 4-8. After standing at room temperature for 30
         min the amount of free ADM released from ADM-OXD was determined by
         gel permeation chromatography.
         Right: ADM-OXD solution (1 mg ADM/ml McIlvain's buffer) was kept
         at 4° for 5 days and free ADM released from ADM-OXD was determined
         by gel permeation chromatography at 24 hr intervals.

incubated with ADM-OXD for less than one hour. The antitumor effect of ADM-
OXD was increased with the treatment time. Furthermore, we have examined
the antitumor spectrum of ADM-OXD against several human cancer cell lines.
ADM-OXD or ADM was incubated with the tumor cells for either two or seven
days. In this experiment ADM-OXD was found to have the same antitumor spectrum against human cancer cell lines as ADM, but the antitumor effect of
ADM-OXD was approximately equal to that of ADM. It is worth mentioning that
ADM-OXD showed stronger cytotoxicity than ADM in some human tumor cell lines
in vitro.

In vivo biological activity. ADM-OXD showed a stronger antitumor effect
than free ADM against rat experimental tumors such as Walker carcinosarcoma
256 and Yoshida sarcoma. On the other hand, ADM-OXD was slightly less cytotoxic than ADM against mouse experimental tumors. Acute toxicity studies
with rats and mice (Ueda et al, 1989), however, showed that ADM-OXD was
about three times less toxic than free ADM. Therefore, we could increase
the dose of ADM-OXD to up to 45 mg/kg, at which the maximum effect of ADM-
OXD was obtained. As shown in Table 5, ADM-OXD showed higher maximum
effects than ADM in mouse experimental tumors regardless of the site of
tumor implantation. ADM-OXD also showed much higher antitumor activity than
ADM against MKN-45 human gastric adenocarcinoma growing in athymic mice.

Biodistribution and excretion. In spite of stronger antitumor activity in
vivo and lower toxicity of ADM-OXD compared to ADM, the therapeutic usefulness was yet to be established. In this regard, we compared the tissue
distribution of ADM-OXD with that of ADM in tumor-bearing rats and mice by
using ^{14}C-ADM. Figure 8 shows the comparative distribution of ADM-OXD and
free ADM in mice bearing Lewis lung carcinoma. Blood level of ADM-OXD was
much higher than that of ADM, indicating that the distribution of the conjugate in the normal tissues might be suppressed. In tumor tissues, ADM-OXD
showed twice as much drug level as free ADM, while the distribution of ADM-

Table 5. Antitumor Effect of ADM-OXD on Rodent Tumors

Mouse tumors

Tumor	Drug	Maximum effect (%)		Dose (mg/kg)
P388	ADM-OXD	ILS	500.0	45
(ip-iv)	ADM		87.5	15
L1210	ADM-OXD	ILS	87.5	45
(ip-iv)	ADM		37.5	5
LLC	ADM-OXD	ILS	114.3	45
(im-iv)	ADM		17.1	15

Rat tumor

Tumor	Drug	ED_{50} (mg/kg)
Walker 256	ADM-OXD	0.67
(im-iv)	ADM	1.43

Male BDF_1 mice (5 weeks) and Wistar strain rats (130–150 g) were inoculated i.m. with 1×10^6 cells of Lewis lung carcinoma and Walker carcinosarcoma 256, respectively. Drugs were injected i.v. five days after the tumor transplantation and animals were sacrificed two weeks after the tumor inoculation to measure the tumor weight. Antitumor effects were expressed as the maximum inhibition ratio in mouse experimental tumor and as ED_{50} in rat experimental tumor. Leukemia L1210 (1×10^5 cells) and leukemia P-388 (1×10^6 cells) were transplanted i.p. to BDF_1 mice. ADM-OXD and ADM were injected i.v. the day after the tumor transplantation. Antitumor effects were expressed as the maximum increase in life span.

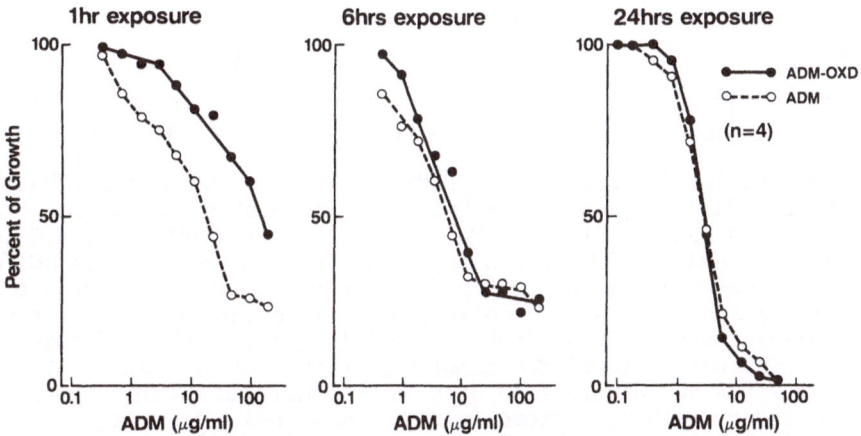

Fig. 7.   Antitumor effect of ADM-OXD against MKN-45 human gastric adenocarcinoma. MKN-45 was inoculated in a 96-well microplate and
incubated in RPMI1640 containing 10% fetal calf serum until the
cell concentration reached about 10^6/well. Then either ADM-OXD or
ADM was added to MKN-45. After culturing with the drug for one,
six and 24 h, MKN-45 was washed with PBS and cultured for another
24 h without the drug. Then viable MKN-45 cells were stained with
crystal violet and cell growth was measured at 590 nm. Each value
represents the average of three samples.

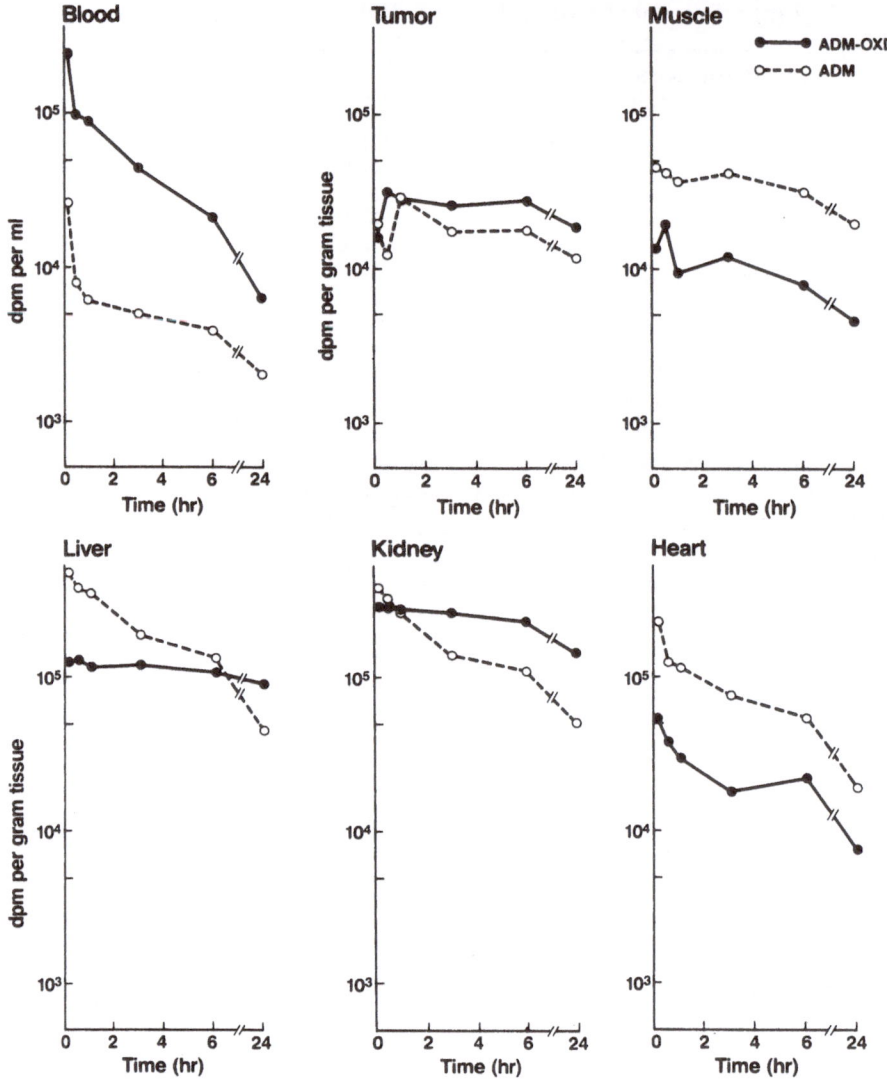

Fig. 8.   Comparative distribution of ADM-OXD and free ADM in tumor bearing
          mice.   Male BDF_1 mice (16-20 g) were inoculated i.m. with 10^6 cells
          of Lewis lung carcinoma.   One week after tumor inoculation C-14-
          labeled ADM or ADM-OXD was injected i.v. at a dose of 0.5 μCi/mouse
          which corresponds to about 3 mg ADM per kg of body weight.   Blood
          and tissue samples were collected at different times after the
          administration of the drugs.   About 100 μl or 100 mg of these
          samples were solubilized completely with 1 ml of Soluene-350
          (Packard) and decolorized with isopropanol and hydrogen peroxide if
          necessary.   Then, radioactivity was measured with a liquid scintil-
          lation counter.   Each value represents the average of three
          samples.

OXD in the heart was lower than that of ADM.   ADM-OXD exhibited a tendency
to be retained longer than free ADM in the RES such as liver.   Rather high
drug levels of ADM-OXD in the kidney indicated that the excretion could take
place there.

   Since the tumor is implanted intramuscularly, specific accumulation of
ADM-OXD into the tumor tissue may be measured by comparing the drug level of

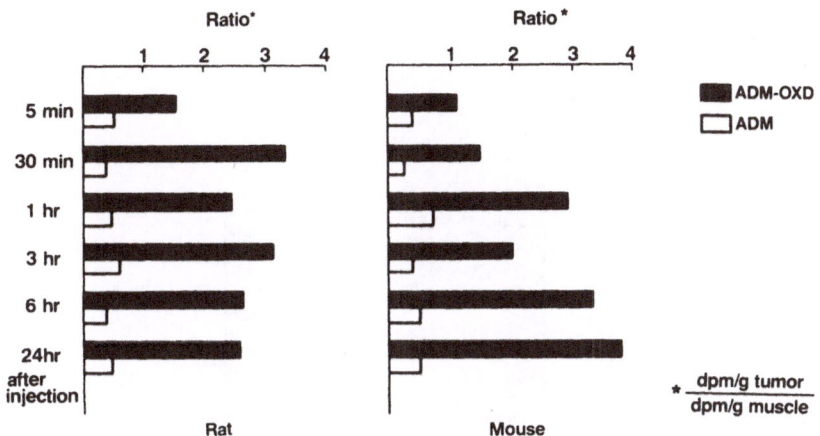

Fig. 9. Tumor/muscle ratio of radioactivity after i.v. injection of radio-
labeled ADM-OXD and ADM. Male Wistar strain rats (100-130 g) and
BDF_1 mice (16-20 g) were inoculated i.m. with 10^6 cells of Walker
carcinosarcoma 256 per rat and 10^6 cells of Lewis lung carcinoma
per mouse, respectively. One week after tumor inoculation, C-14-
labeled ADM-OXD or ADM was injected i.v. at a dose of 0.5
μCi/animal (ca. 3 mg ADM/kg). Radioactivities in the tumor and
normal muscle were measured as described in Fig. 8. Tumor/muscle
ratio was calculated by dividing the radioactivity of 1 g of tumor
by that of 1 g of normal muscle in both rats and mice. Each value
is the mean of three samples.

the tumor tissue with that of the normal muscle. The tumor/muscle ratio of
free ADM was less than 1.0 throughout the experimental period (Fig. 9)
whereas that of ADM-OXD ranged from about 2.0 to 4.0 in both rats and mice.
This result strongly indicates that ADM-OXD accumulates into the tumor tissues more specifically than free ADM, meeting the requirements for passive
targeting of ADM. As shown in Fig. 10, approximately 50% of ADM-OXD was
excreted into the urine by six h and 15% into the feces by 24 h after administration. On the other hand, only 12% of the drug was excreted into the
urine by 24 h in the case of free ADM. This is in sharp contrast to the
finding that biliary excretion is the main route for free ADM (Tavoloni and
Guarino, 1980), and indicates a different fate of ADM-OXD in vivo compared
to ADM.

Our research has made it clear that the conjugation of ADM to oxidized
dextran results in the drastic alteration of characteristics of ADM, enabling its passive targeting.

Antibody-Adriamycin Conjugate

The use of monoclonal antibodies (MoAbs) as a DDS has attracted our
attention for the last decade, because of the great affinity of antibodies
to antigens on the cell surface (e.g. Pimm, 1988). MoAbs, especially those
to tumor cells, are thus being exploited in therapy for the targeting of
toxins to the tumor site. For instance, there have been several reports on
the chemical conjugation of an anticancer drug to a MoAb for the "active
targeting" of the drug (Tsukada et al, 1982; Manabe et al, 1985; Dillman et
al, 1986; Tsukada et al, 1987; Johnson and Laguzza, 1987). Major questions
associated with such derivatives are immunogenicity, whether the conjugate
survives in the circulation and retains the same antibody titer as the
intact MoAb, and availability of MoAbs in kilogram quantities (e.g. Pimm,

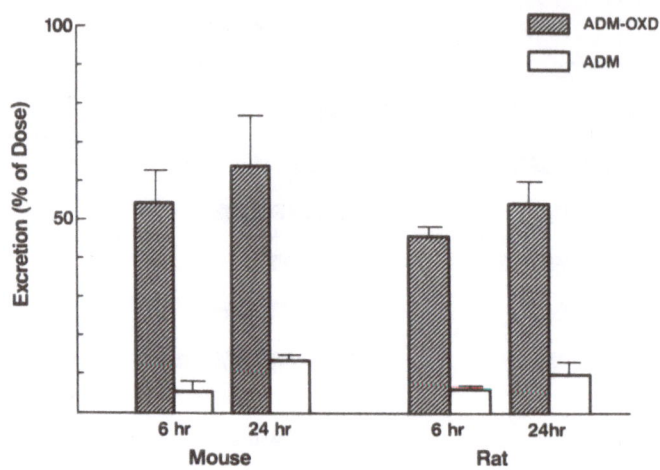

Fig. 10. Urinary excretion of ADM-OXD and free ADM. ^{14}C-labeled ADM-OXD and ADM were injected into tumor bearing rats and mice as described in Fig. 8. Urine was collected during 6 and 24 hours after i.v. administration of the drugs. The radioactivity of urine was measured after solubilizing with Soluene-350 (Packard) as described in Fig. 8.

1988). Having made progress in our study with ADM-OXD, we have also become interested in the development of a MoAb conjugate employing the ADM-OXD as a spacer, aiming at more efficient targeting.

We have recently found a mouse MoAb against human gastric adenocarcinoma MKN-45 (KM10) which reacts with tumors of digestive organs such as colon, stomach, Pappila Vater and pancreas (Ohyanagi et al, 1988). KM10 also reacts with normal tissues (colon, stomach, small intestine and esophagus), but the reaction is restricted to the luminal surface of the columna cells, suggesting that KM10 is not absolutely specific against carcinoma and might show a distribution comparable to carcinoembryonic antigen. In general, mouse antitumor MoAbs have been used for targeting therapy because of their efficient production rate and easy cloning. However, immunogenicity for the mouse protein is an important problem in the clinic. To overcome this obstacle, chimeric MoAb has been widely studied and recent progress in hybridoma technology and genetic engineering has already made it feasible in our hands. We have recently succeeded in manufacturing the chimeric KM10 on an industrial scale with a high expression system and an effective mass culture method, thus allowing the study of KM10.

Preparation of KM10-ADM conjugate. KM10-ADM conjugate was derived from ADM-OXD which was prepared exactly as described above for ADM-OXD. thus, to a 50 mM phosphate buffer solution of KM10 (4% (w/v), pH 7.5) a phosphate buffer solution of ADM-OXD was added while stirring under cooling. The entire mixture was stirred at 20-25°C for 18 h, then subjected to a Sephacryl S300 côlumn chromatography to give a fraction of KM10-ADM conjugate in 80-90% yield. The molecular weight of the conjugate was estimated by gel permeation chromatography to be 230 kd and the binding ratio of ADM to KM10 around seven. The antibody titer of the conjugate was estimated by ELISA to be about 75% of that of intact KM10 (Ueda et al, 1989; Ohyanagi et al, 1988).

Effect on MKN-45 xenograft in nude mice. Table 6 shows the antitumor effect of KM10-ADM conjugate on MKN45 xenograft in nude mice as compared with that

Table 6. Effect of KM10-Adriamycin Conjugate on MKN-45 Xenograft in Nude Mice

Agent	Dose (mg ADM/kg/day)	Inhibition Rate (%)		
		10th	30th	50th day
Conjugate	5	26.5	58.5	75.9
	15	42.7	74.1	87.6
Free Adriamycin	5	2.8	35.2	43.6
	15	toxic	—	—

Nude mice were inoculated subcutaneously with 1×10^7 cells per mouse. When the tumor developed to a mass of about 8 mm in diameter, the mice were divided into four groups, and the conjugate or free ADM was injected intravenously through tail vein at doses of five and 15 m/kg/day as ADM three times at three day intervals.

of ADM. The effect of the KM10 conjugate was markedly higher than that of ADM. At a dose of 5 mg/kg/day, the inhibition rate of the conjugate on day 50 was 75.9% while that of ADM was only 43.6%. In the 15 mg dosing group, the conjugate showed a strong antitumor effect with 87.6% inhibition rate on day 50 without death. Free ADM showed its inherent toxicity resulting in the death of all animals even on day 10. These results encourage us to pursue further development of the KM10 conjugate. Preclinical studies are now in progress.

CONCLUSIONS

A great deal of attention has been directed to prolonging the circulatory time of drugs or to delivering drugs efficiently into specific sites. For these reasons different approaches involving modification of the drug itself or optimization of its formulation have been followed. We dealt with the improvement of drugs as formulations based on the concept of DDS. The use of fat emulsions as the carrier indicated that prostaglandin E_1 and dexamethasone incorporated into the emulsion system are localized to inflamed tissues or vascular walls, thus providing clinical safety as well as higher effectiveness.

ADM-OXD appears to show interesting properties such as different in vivo fate from free ADM and to have higher effectiveness and lower toxicity than free ADM, suggesting promising results in clinical uses. The use of monoclonal antibodies (MoAbs) against tumor tissues may become of value for the treatment of cancer patients. In terms of the practical development of anticancer agents linked to MoAb, the production of chimeric MoAbs on an industrial scale is essential. In this regard, our ADM-OXD-KM10 conjugate also seems promising.

It is clear that there has been great progress in the area of DDS but that much work remains to be performed. The availability of safe and effective preparations which can reach disease sites specifically will provide us with opportunities for exciting applications. Thus, constant efforts must be made to ensure the production of reliable preparations. Responsibility for anything concerning the drugs must be recognized as "Adiutor pro Arte Medicina" (a supporter for medical sciences).

Acknowledgements

Many thanks are due to H. Okamoto, M. Watanabe, T. Hamano, T. Imagawa, T. Inoue, K. Munechika and Y. Kanoh, all at the Research Division of the Green Cross Corporation, for their valuable work on the study of DDS.

REFERENCES

Davis, S.S., Hadgraft, J. and Palin, K.J., 1985, Pharmaceutical emulsions, in: "Encyclopedia of Emulsion Technology", P. Becher, ed., Dekker, New York.

Dillman, R.O., Shawler, D.L., Johnson, D.E., Meyer, D.L., Koziol, J.A. and Frincke, J.M., 1986, Preclinical trials with combination and conjugates of T101 monoclonal antibody and doxorubicin, Cancer Res., 46:4886.

Esumi, Y., Mitsugi, K., Sekine, S., Yokoshima, T., Kawano, Y., Suzuki, A. and Suwa, T., 1986, Absorption, distribution, metabolism and excretion of ^{3}H-Lipo PGE1, Kiso to Rinshou, 20:161.

Hamano, T., Sintome, M. and Watanabe, M., 1986, Vasodilation activity of Lipo-PGE1, Kiso to Rinshou, 20:93.

Hoshi, K., Mizushima, Y., Shiokawa, Y., Kageyama, T., Honma, M., Kashiwazaki, K., Tsunematsu, T. and Kaneko, K., 1985, Double-blind study with liposteroid in rheumatoid arthritis, Drug Exptl.Clin.Res., XI:621.

Ii, S., Okamoto, H., Yokoyama, K. and Suyama, T., 1988, Comparative pharmacokinetic study of dexamethasone palmitate and dexamethasone disodium phosphate in chronic rheumatoid arthritis patients, Yakuri to Chiryou, 16:865.

Iwai, K., Maeda, H. and Konno, T., 1984, Use of oily contrast medium for selective drug targeting to tumor: enhanced therapeutic effect and x-ray image, Cancer Res., 44:2115.

Johnson, D.A. and Laguzza, B.C., 1987, Antitumor xenograft activity with a conjugate of Vinca derivative and the squamous carcinoma-reactive monoclonal antibody PF1/D, Cancer Res., 47:3118.

Kaneko, K., 1985, Studies on liposteroid therapy for rheumatoid arthritis, Ensho, 5:241.

Manabe, Y., Tsubota, T., Haruta, Y., Kataoka, K., Okazaki, M., Haisa, S., Nakamura, K. and Kimura, I., 1985, Production of a monoclonal antibody-mitomycin C conjugate, utilizing dextran T-40 and its biological activity, Biochem.Pharmacol., 34:289.

Martin, M.F.R., 1980, Prostaglandin E$_1$ in the treatment of systemic sclerosis, Ann.Rheum.Dis., 39:44.

Miura, K., Kumon, S. and Yamao, F., 1982, Phase I study of dexamethasone palmitate incorporated in a fat emulsion (Liposteroid), Kiso to Rinshou, 16:5928.

Mizushima, Y., Hamano, T. and Yokoyama, K., 1982, Tissue distribution and anti-inflammatory activity of corticosteroids incorporated in lipid emulsions, Ann.Rheum.Dis., 41:263.

Mizushima, Y., Yanagawa, A. and Hoshi, K., 1983, Prostaglandin E$_1$ is more effective when incorporated in lipid microspheres for treatment of peripheral vascular diseases in man, J.Pharm.Pharmacol., 35:666.

Momma, K., 1984, Prostaglandin E$_1$ treatment of ductus dependent infants with congenital heart disease, Intern.Angiol., 3(Suppl.):33.

Nakura, K., Hamano, T., Sintome, M. and Watanabe, M., 1986, Location of lipo PGE1 in blood vessel, Kiso to Rinshou, 20:143.

Ohyanagi, H., Ishida, H., Ishida, T., Soyama, N., Yamamoto, M., Okumura, S., Kano, Y., Ueda, Y. and Saitoh, Y., 1988, A monoclonal antibody, KMO 10 reactive with human gastrointestinal cancer and its application for immunotherapy, Jpn.J.Cancer Res.(Gann), 79:1349.

Okamoto, H., Watanabe, M., Ohyanagi, H. and Saitoh, Y., 1985, In vitro
hydrolysis of dexamethasone-21-palmitate, Ensho, 5:219.

Otomo, S., Mizushima, Y., Aihara, H., Yokoyama, K., Watanabe, M. and
Yanagawa, A., 1985, Prostaglandin E_1 incorporated in lipid microspheres (Lipo PGE1), Drugs Exptl.Clin.Res., X:627.

Pimm, M.V., 1988, Drug-monoclonal antibody conjugates for cancer therapy:
potentials and limitations, Crit.Rev.Ther.Drug Carrier Syst., 5:189.

Shaw, I.H., Knight, C.G., Thomas, D.P.P., Phillips, N.C. and Dingle, J.T.,
1979, Liposome-incorporated corticosteroids: 1. The interaction of
liposomal cortisol palmitate with inflammatory synovial membrane,
Br.J.Exp.Pathol., 38:553.

Sugai, T., Okamoto, H. and Yokoyama, K., 1984, Vasoconstrictor activity and
percutaneous absorption of dexamethasone-21-palmitate, a main component of "Liposteroid", Hifu, 26:525.

Tavoloni, N. and Guarino, A.M., 1980, Disposition and metabolism of adriamycin in rats, Pharmacology, 21:244.

Tomlinson, E., 1986, (Patho)physiology and the temporal and spatial aspects
of drug delivery, in: "Site-specific Drug Delivery", E. Tomlinson and
S.S. Davis, eds., John Wiley and Sons Ltd., Chichester.

Tsukada, Y., Hurwitz, E., Kashi, R., Sela, M., Hibi, N., Hara, A. and Hirai,
H., 1982, Chemotherapy by intravenous administration of conjugates of
daunomycin with monoclonal and conventional anti-rat -fetoprotein
antibodies, Proc.Natl.Acad.Sci.USA, 79:7896.

Tsukada, Y., Ohkawa, K. and Hibi, N., 1987, Therapeutic effect of treatment
with polyclonal or monoclonal antibodies to -fetoprotein that have
been conjugated to daunomycin via a dextran bridge: studies with an
-fetoprotein-producing rat hepatoma tumor model, Cancer Res.,
47:4293.

Ueda, Y., Munechika, K., Kikukawa. A., Kano, Y., Yamanouchi, K. and
Yokoyama, K., 1989, Comparison of efficacy, toxicity and pharmacokinetics of free adriamycin and adriamycin linked to oxidized dextran
in rats, Chem.Pharm.Bull., 37:1639.

Yokoyama, K., Okamoto, H., Watanabe, M., Suyama, T. and Mizushima, Y., 1985,
Development of a corticosteroid incorporated in lipid microspheres
(Liposteroid), Drug Exptl.Clin.Res., X:611.

CONTROLLED DELIVERY TO THE BRAIN

Henry Brem

Departments of Neurological Surgery, Ophthalmology and
Oncology, Johns Hopkins University School of Medicine
Baltimore, Maryland, USA

INTRODUCTION

Controlled delivery of drugs directly to the site of a growing brain
tumor has the potential to increase drug efficacy while reducing systemic
side effects. Hochberg and Pruitt (1980) showed that 90% of malignant
gliomas recur within a 2 cm margin of the original tumor. Thus, targeting
therapy in this region holds the promise of controlling tumor growth and
minimizing the systemic toxicity of the drugs. Localized, prolonged delivery of drugs to the brain is now possible by using polymeric delivery systems.

Many attempts have been made to increase the delivery of chemotherapeutic drugs to the brain tumor site, including high-dose systemic treatment
with bone marrow rescue techniques (Greig, 1987; Mortimer et al,1973),
selective intracranial arterial infusion (Greenberg et al, 1984), transient
osmotic disruption of the blood brain barrier (Fishman, 1987; Neuwelt et al,
1986; Warnka et al, 1987), cerebrospinal fluid perfusion (Blasberg, 1977;
Blasberg et al, 1987), non-biodegradable polymers (Rosenblum et al, 1973)
and direct infusion into a brain tumor utilizing catheters (Garfield et al,
1973; Ringkjob, 1968; Rubin et al, 1966; Weiss et al, 1969). Several factors may contribute to the failure of the direct infusions: the drug used
for infusion studies, methotrexate, may be inadequate for treating gliomas
(Bouvier et al,1987), limited infusion distance by use of catheters
(Blasberg, 1977; Blasberg et al, 1987; Harbough et al, 1988), or that the
generally used intermittent infusions (usually weekly) do not allow a sustained diffusion gradient (Levin et al, 1980). In addition, the clinical
use of catheters is complicated by high rates of infection, of obstruction
and malfunction due to clogging by tissue debris, and unpredictable release
rates.

To overcome these difficulties, and yet to achieve the goal of by-
passing the blood-brain barrier, thereby increasing the number of drugs
available, and their local concentrations while minimizing the systemic side
effects, we began to explore the role of biodegradable polymer delivery

Targeting of Drugs, Edited by G. Gregoriadis *et al.*
Plenum Press, New York, 1990

systems. Before 1973, there were no chemical methods of prolonged control-led release of molecules greater than molecular weight of 500 that were biocompatible in vivo. Polyacrylamide gels were capable of macromolecular release but were highly inflammatory (Gimbrone et al, 1974). In 1976, Langer and Folkman reported the release kinetics of macromolecules from ethylene-vinyl acetate (EVAc) and several other polymers. Subsequently, Langer, Brem and Tapper (1981) reported the biocompatibility of this non-degradable polymer, EVAc, in the rabbit cornea. Extensive biological and clinical applications have been developed for the EVAc polymer including contraception, insulin delivery, cancer chemotherapy, glaucoma treatment, dental caries prevention, and asthma therapy (Langer and Wise, 1986). However, the potential of polymeric controlled delivery for neural applic-ations had not been previously explored. Considering the low bioavailabil-ity of many therapeutic agents due to the blood-brain barrier, a successful sustained release system to the brain would represent a significant achievement in chemotherapy.

The EVAc delivery system is predictable and reproducible (Langer, 1983; Langer and Wise, 1986). The limitations of the system are its non-bio-degradability, its hydrophilicity which may not protect certain unstable drugs (such as BCNU), and that it releases by bulk erosion which could lead to an uncontrolled burst of chemotherapeutic drug in the brain, causing serious toxicity.

By contrast, the newly developed (poly[bis(p-carboxyphenoxy)propane-sebacic acid] (PCPP-SA) has the following advantages: it is biodegradable over periods ranging from days to years depending on its formulation, it is hydrophobic thereby protecting unstable compounds from the surrounding environment, and it degrades by pure surface erosion which therefore approaches zero order kinetics for drug release. Theoretically PCPP-SA could be made available in a variety of forms such as sheets, rods, micro-spheres, or wafers, and should be able to be constructed in a wide range of physical forms ranging from rigid to very flexible. Incorporation of drugs into the PCPP polymer is very simple, requiring no solvents, and is accomp-lished at low temperatures.

Leong et al (1986) have tested PCPP-SA polymer and its breakdown pro-ducts, and shown them to be non-mutagenic, non-cytotoxic and non-terato-genic. Endothelial cell and smooth muscle cell growth in tissue culture was unaffected by being plated on a layer of polyanhydride polymer.

Many chemotherapeutic agents have short half-lives in vitro (e.g. BCNU degrades to an inactive form in about 15 minutes in serum). Since polymers such as EVAc do not protect the drug from the surrounding environment, it is possible that the drug is decomposed inside the EVAc matrix before it can be released. A matrix-degradation release mechanism, as achieved with polyan-hydride, is therefore desirable for unstable drugs. The hydrolytic degrad-ation mechanism of the polyanhydrides make them ideal for administering chemotherapeutic agents. The polymer reacts with water to form dicarboxylic acids. By contrast biodegradation of some other potential release agents such as polyesters or polyamides are affected by esterases or amidases which may be impaired by chemotherapeutic agents, thereby altering the predicted matrix characteristics.

In order to determine if this biodegradable controlled release polymer could be utilized to treat human brain tumors we developed a series of experiments to explore (i) the compatibility of this polymer with neural tissue and (ii) the efficacy of the controlled release of a well character-ized chemotherapeutic agent (BCNU) incorporated in this polymer in inhibit-ing the growth of a malignant glioma. The preliminary results from these studies are described below.

BIOCOMPATIBILITY OF PCPP-SA

Rabbit Cornea

We tested the biocompatibility of the biodegradable polyanhydride, poly[bis(p-carboxyphenoxy)propane-sebacic acid] copolymer (PCPP-SA) in the rabbit cornea, the rabbit brain, the rat brain, and the monkey brain.

The PCPP-SA was initially tested in the rabbit cornea bioassay (Langer et al, 1981), a highly sensitive system in which inflammatory effects of an implant can be easily identified. These tests showed that the polymer was highly biocompatible and therefore suitable for further definitive testing in the brain of several species (Leong et al, 1986). None of the corneas had an inflammatory response (corneal edema or neovascularization) and all remained clear throughout the 3 weeks of observation. Thus, we proceeded with definitive biocompatibility testing of PCPP-SA in the rat, rabbit and monkey brain.

Rat Brain

In rat brain, we compared the tissue reaction in the presence of PCPP-SA to that seeen with absorbable gelatin sponge (Gelfoam) and with oxidized regenerated cellulose (Surgicel) (Tamargo et al, 1989; Blasberg, 1977). None of the animals showed any behavioral changes or neurological deficits suggesting either systemic or localized toxicity from the biodegradable polyanhydride. All rats survived to the scheduled date of sacrifice. At day 3, all three implants evoked a well demarcated, acute inflammatory response, marked in the case of PCPP-SA and Surgicel, and milder in the case of Gelfoam. Lymphocytes were first seen on day 10, indicating the transition to a chronic inflammatory reaction which resolved with the total degradation of PCPP-SA by day 28 and of Gelfoam by day 36, but persisted to day 36 with Surgicel. In summary, PCPP-SA evoked a well localized inflammatory reaction, comparable to that of Surgicel which resolved as the polymer degraded over one month.

Rabbit Brain

We then determined the biocompatibility of PCPP-SA in the rabbit brain (Brem et al, 1989). Twenty adult New Zealand white male rabbits underwent implantation of PCPP-SA in one frontal lobe and absorbable gelatin sponge (Gelfoam) in the other frontal lobe. As we recently reported (Brem et al, 1989a), all of the animals that received implants survived to their date of sacrifice and none of the animals showed behavioral changes or other neurological deficits suggestive of toxicity. Histological examination showed no significant differences between the tissue reaction in the presence of PCPP-SA or that in the presence of Gelfoam. We concluded that PCPP-SA was nontoxic and biocompatible in the rabbit brain.

Monkey Brain

Since we had established the biocompatibility of PCPP-SA in rat and rabbit brain, we tested the effects in monkey brain, with the addition of a study of polymer containing BCNU (Brem et al, 1988b). Fifteen adult male Cynomologous monkeys were randomized into one of three groups of five monkeys: All animals underwent a right frontal lobectomy. Group 1 received 660 mg PCPP-SA discs 1.9% loaded with 12.5 mg of BCNU; Group 2 received empty discs; and Group 3 had only lobectomy but no implant. Post-operatively, the animals were evaluated daily for any evidence of systemic toxicity or neurological changes. Complete blood counts and serum analyses were obtained on the day of surgery and on (postoperative) days 2-5, 8-10, 12-14, 16-18, 40-42 and 71-78. Three monkeys in each group were sacrificed on days

16-18 and the other two were sacrificed on days 71-81. Complete autopsies were carried out in all the animals. There were neither adverse systemic or neurological changes, nor differences in blood chemistries or hematological studies. In brief, no adverse clinical or histological effects were seen with any of the treatment groups.

We also assessed (Brem et al, 1988a) the radiological features of PCPP-SA in the monkey brain by computerized tomography (CT) and magnetic resonance imaging (MRI). CT scans (with and without contrast) were performed on all monkeys on days 8-12 and on selected monkeys on days 72-79. MRI scans were performed on 12/15 monkeys on days 12-30. Three monkeys in each group were sacrificed on days 16-18 and the other two were sacrificed on days 71-81. Complete autopsies were carried out in all the animals. CT and MRI images of an isolated polymer in vitro were also obtained. In early CT imaging, the polymer appeared as a radiopaque density. There was no evidence of increased edema or mass effect in the brain surrounding either the empty polymer or the polymer with BCNU compared to the control group. In later CT imaging, there was no evidence of the polymer and the post-operative changes seen on the earlier scans had resolved. In the MRI study, the polyanhydride was identified as a hypointense structure surrounded by iso- or hyperintense structures consistent with edematous brain or, in some cases, blood. Thus both the polymer and the brain tissue reaction can be evaluated with both CT and MRI.

In conclusion, there was no systemic toxicity from the implants in the monkey brain. All of the monkeys survived to the scheduled sacrifice day. The blood tests were normal throughout. The CT and MRI scans localized the degrading polymers and showed minimal reaction to them. Autopsies were remarkable only for localized reaction around the polymer (Brem et al, 1988b). There were no adverse effects of PCPP-SA in any of the animal studies. We next investigated pharmacokinetics of the localized, controlled release of the best characterized chemotherapeutic agent, BCNU, against brain tumors.

RELEASE KINETICS OF BCNU FROM POLYMER IMPLANTED IN THE BRAIN

Release of BCNU in the Rat

We initially described (Yang et al, 1989) the release kinetics of BCNU released from polymers in the rat. Polymer cylinders loaded with BCNU were implanted intraperitoneally in 34 Fischer 344 rats. Groups of four rats were exsanguinated by direct cardiac puncture 1, 4, 12 and 24 hours and 2, 3, 4, 5 and 6 days (only 2 rats on day 6) after implantation. The concentration of BCNU in whole blood was determined by the Bratton-Marshall assay.

We found that BCNU was released continuously into the circulation throughout the duration of the experiment. The concentration of BCNU in whole blood dropped from 3.5 µg/ml at 1 hour to 0.3-0.5 µg/ml by 24 hours, staying in this range for the 6 days of the experiment. The sensitivity limit of our assay necessitated a toxic dose that killed the rats by the sixth day. In vitro release data obtained simultaneously in phosphate buffer suggested that the polymers would continue to release BCNU for at least 1 week.

Using the same technique, we have defined the release kinetics of BCNU in the rat brain (Yang et al, 1989). Polymers containing BCNU were implanted in the brain of rats and the rats were serially sacrificed. Brain homogenates were assayed for the presence of BCNU by the Bratton-Marshall assay. As in the periphery, BCNU is released in high concentrations for an extended period of at least two weeks in the brain, but it saturates the tissues at much higher concentrations than those in the serum of the rats with peritoneal implants.

AUTORADIOGRAPHY OF RADIOLABELED DRUG

BCNU Release in the Rabbit

To understand better the pharmacokinetics of BCNU released from the polymer in the brain, we studied the distribution of tritium-labelled BCNU released from PCPP-SA in the rabbit brain by performing autoradiographic analysis of drug distribution within the brain, and compared these results to brain drug distribution obtained by the direct injection of labelled BCNU into the brain. PCPP-SA cylinders containing three different concentrations of [^{3}H] BCNU, labelled in the chloroethyl methylene groups, were implanted into New Zealand white rabbits. An additional group receiving [^{3}H]-inulin loaded in polyanhydride was used as a control. Two rabbits from each group were sacrificed on each of days 3, 7, 14 and 21 following surgery. Brains were frozen and processed for sectioning. After every 28 sections, the following two sections were saved for autoradiography and stained with hematoxylin and eosin. Exposure was for two weeks. The autoradiographic image was digitized and quantitatively analyzed by use of a computer-assisted image analysis system. Hematoxylin and eosin-stained specimens were microscopically examined for any remaining polymer and for any tissue reaction.

The autoradiographs showed that by the third day radioactivity associated with BCNU was dispersed throughout the ipsilateral hemisphere, and some was even found in the contralateral hemisphere. Subsequently, there was a decrease in the area of brain which showed significant levels of radioactivity associated with BCNU. Twenty one days after implantation, the brain tissue immediately surrounding the degrading implant showed drug concentrations two standard deviations above the background. By contrast, such levels of drug activity were no longer detectable 48 hours after direct injection. Corresponding punch biopsies and immediate HPLC analysis confirmed the correlation between the radioactive label and intact BCNU. We also determined that the concentration adjacent to the polymer was 6.5 mM and, even at a distance of 10 mm, the concentration was 200 µM. These concentrations are much higher than achieved with standard intravenous administration (Chasin et al, 1989).

The three concentrations of BCNU used in this study corresponded to 42.5, 85 and 170 µg/mm^2 of brain surface. Following histological examination of the brain slices, it was concluded that there were no significant differences between the brain reaction to any of the three groups. A small rim of necrosis and a mild inflammatory response was found around the rim of the polymer at each of the three doses of BCNU.

EFFECTIVENESS OF PCPP-SA LOADED WITH BCNU

With the biocompatability of the BCNU-containing polymers in the brain established and having gained insight into the drug distribution kinetics, we then explored the effectiveness of this approach as compared to standard experimental methods for treating brain tumors.

9L Gliosarcoma Model

We tested (Brem et al, 1989b; Tamargo et al, 1988a) the efficacy of the sustained release of BCNU in controlling the growth of the malignant glioma and compared this technique to the standard systemic administration of BCNU.

The 9L gliosarcoma was implanted in the flank of sixty Fischer 344 rats. BCNU (3.8 mg) was incorporated into polymers (15 mg). The rats were assigned to one of six treatment groups of 10 rats: (i) implantation of a BCNU-loaded (18 mg/kg) polymer adjacent to the tumor; (ii) implantation of a BCNU-loaded (18 mg/kg) polymer in the contralateral flank; (iii) intraperitoneal administration of BCNU 13 mg/kg; (iv) intraperitoneal administration of BCNU 18 mg/kg; (v) implantation of an empty polymer adjacent to the tumor; and (vi) sham manipulation of the tumor. Treatment was initiated at a mean tumor volume of 320 mm^3.

The localized, sustained release of BCNU from the polymer at the tumor site resulted in the longest growth delay in comparison with the controls. Whereas the intraperitoneal administration of BCNU 18 mg/kg, and the implantation of the BCNU-loaded polymer in the contralateral flank delayed growth for 9-14 days, the implantation of the BCNU-loaded polymer adjacent to the tumor delayed growth for 22 days.

This study suggested that BCNU released from a polymer was biologically active and that the regional, continuous release of BCNU directly at the tumor site from an implantable polymer may be a more efficacious way of treating malignant gliomas.

Intracranial Efficacy of BCNU PCPP-SA against Simultaneous Tumor Implant

We then proceeded to test whether we could similarly control the 9L gliosarcoma growing in the brain of the rats by simultaneously implanting tumor pieces and BCNU-loaded polymers intracranially. Sixty adult male Fischer 344 rats were randomized into one of six groups; (i) tumor plus polymer with BCNU; (ii) tumor plus polymer without BCNU; (iii) tumor alone; (iv) polymer with BCNU alone; (v) polymer without BCNU alone; and (vi) no tumor and no polymer. The animals underwent craniotomies for the implantation of the tumor and of PCPP-SA cylinders. The animals were examined twice daily for any signs or symptoms of neurological or systemic deterioration and followed until the time of death. The animals bearing tumor alone or tumor with a blank polymer died from their tumors between post-operative days 10 and 26. By contrast, in the group of animals with tumor and BCNU-loaded polymer, three rats died between days 42 and 56 and seven rats lived to 150 days, at which time all survivors were sacrificed. There was no histological evidence of tumor in any of the brains of long term survivors implanted with BCNU in the polymer and tumor.

Efficacy of BCNU and PCPP:SA against an Established Tumor

These results led us to a similar experiment in which already-established intracranial tumors were reoperated and treated with BCNU-loaded polymers. Fifty-four rats were studied and divided into six groups. 9L gliomas or sham implantations were performed on the first day. On the fourth post-operative day, BCNU polymer or empty polymer was implanted at the tumor site and intraperitoneal injection by either BCNU or ethanol and normal saline was administered, according to the protocol shown in Table 1.

We found that with empty polymer, i.e. the untreated controls, the average survival was 11 days, with intraperitoneal BCNU the average survival was 27 days, and with intracranial BCNU and PCPP-SA the average survival was 62 days. In addition, two of the nine rats with intracranial BCNU in PCPP-SA survived to the time of planned sacrifice, 120 days, with no evidence of tumor; whereas none of the rats with empty polymers or intraperitoneal BCNU injections survived beyond 58 days (Brem et al, 1989b; Tamargo et al, 1988b). Thus polymer implantations caused 'cures' which were not seen with other modes of BCNU delivery.

160

Table 1. Effect of BCNU Polymer in Rats Implanted with Gliosarcoma

| Group | Implantation with 9L Glioma | Brain Implant | | Intraperitoneal | | Mean Survival (days) |
		Empty Polymer	BCNU Polymer	BCNU	Ethanol & Normal Saline	
1	−	+	−	−	+	120
2	−	+	−	+	−	63
3	−	−	+	−	−	93
4	+	+	−	+	−	27
5	+	−	+	−	+	62
6	+	+	−	−	+	11

In conclusion, therefore, in this series of studies we established that: (i) PCPP-SA was biocompatible and could be implanted safely in the brains of rodents and primates and (ii) BCNU incorporated in PCPP-SA could be released in a sustained, controlled fashion. This form of delivery was shown to effectively inhibit the growth of an experimental malignant glioma, proving to be superior to the standard systemic administration of BCNU.

CLINICAL TRIALS

The Food and Drug Administration reviewed the above data and concluded in 1987 that we could proceed with a Phase I Clinical Trial of PCPP-SA (20:80) containing BCNU. A Phase I clinical study was started to investigate the safety of the BCNU polyanhydride (PCPP-SA 20:80) implanted at the site of tumor resection (Brem et al, 1989c). This clinical study was based on the known tendency of malignant brain tumors to recur locally and the fact that BCNU given systemically has a half-life of only 12 minutes (Loo and Dion, 1965). In this study, the BCNU is protected from degradation within the anhydrous polymer. After the tumor was extirpated, we covered the surgical cavity with BCNU-PCPP-SA so that the target tumor cells would be exposed directly to the drug. This study included patients who had previous surgery for Grade III or IV gliomas confirmed by histology and previous definitive external beam radiation.

Twenty one patients were studied at five institutions in this clinical trial. Five patients were entered with a 1.93% BCNU loading dose and a surface concentration of 25 µg per mm^2. Once we determined that this was a safe dose, we enrolled an additional five patients with 3.85% BCNU loading and a surface concentration of 50 µg per mm^2. Four months after these patients were entered, 11 patients were enrolled with 6.37% BCNU loading in an 82.5 µg per mm^2 concentration. The results of this study are currently being analyzed. In the eight patients investigated at Johns Hopkins Hospital, we found that the polymers were well tolerated and did not cause any detectable toxicity. We were surprised to find that two to three months after implantation, contrast CT scans showed a ring enhancement around the polymers with mild mass effect that gradually resolved. As of June 8, 1989, the average survival after polymer implantation of our eight patients is 45 weeks and the average for the three survivors is 68 weeks. Overall, seven of the 21 patients are still alive, with an average for all of the survivors of 67 weeks. This is a significant improvement in the expected survival (14

weeks) with recurrent glioblastoma. The Phase I study showed that the polymer utilized is safe as predicted from our experimental work.

To determine the effectiveness of this treatment approach, we have now
initiated a prospective randomized placebo-controlled double-blinded trial.
This study will be carried out at 16 institutions over a one year period and
involve about 250 patients. Patients who have failed conventional treatment
will either receive PCPP-SA with BCNU (3.85% loading and a surface concentration of 50 μg per mm^2) or PCPP-SA discs without chemotherapeutic drug.
This clinical study should enable us to determine the effectiveness of BCNU
impregnated polymers in treating recurrent malignant brain tumors.

It is our hope that through further experimentation we will be able to
address the many variables inherent in our initial studies and learn how
best to apply the techniques of controlled delivery to better understand and
treat brain tumors.

REFERENCES

Blasberg, R.G., 1977, Methotrexate, cytosine arabinoside, and BCNU concen-
tration in brain after ventriculocisternal perfusion, Cancer Treat.
Rep., 61:625
Blasberg, R.G., Patlak, C.S. and Shapiro, W.R., 1977, Distribution of methotrexate in the cerebrospinal fluid and brain after intraventricular
administration, Cancer Treat.Rep., 61:633.
Bouvier, G., Penn, R.S., Kroin, J.S., Beique, R. and Guerard, M.J., 1987,
Direct delivery of medication into a brain tumor through multiple
chronically implanted catheters, Neurosurgery, 20:286.
Brem, H., Ahn, H., Tamargo, R.J., Pinn, M.L. and Chasin, M., 1988, A bio-
degradable polymer for intracranial drug delivery: A radiological
study in primates. American Association of Neurological Surgeons,
Toronto, p. 349.
Brem, H., Kader, A., Epstein, J.I., Tamargo, R., Domb, A., Langer, R. and
Leong, K., 1989, Biocompatibility of bioerodible controlled release
polymers in the rabbit brain. Selective Cancer Therapeutics, 5:55.
Brem, H., Tamargo, R.J. and Olivi, A., 1990, Delivery of drugs to the brain
by use of a sustained release polymer system, in: "New Technologies
and Concepts for Reducing Drug Toxicity", H. Salem, ed., Telford
Press, Caldwell, N.J., in press.
Brem, H., Tamargo, R.J., Pinn, M. and Chasin, M., 1989, Biocompatibility of
a BCNU-loaded biodegradable polymer: A toxicity study in primates.
American Association of Neurological Surgeons, Toronto, p. 381.
Chasin, M., Domb, A., Ron, E., Mathiowitz, E., Leong, K., Laurencin, C.,
Brem, H. and Langer, R., 1990, Polyanhydrides as drug delivery
systems, in "Biodegradable Polymers as Drug Delivery Systems",
R. Langer and M. Chasin, eds., Marcel Dekker Inc., New York, in
press
Fishman, R.A., 1987, Is there a therapeutic role for osmotic breaching of
the blood-brain barrier? Ann. Neurol., 22:298.
Garfield, J., Chir, M. and Dayan, A.D., 1973, Postoperative intracavitary
chemotherapy of malignant gliomas, J.Neurosurg., 39:315.
Gimbrone, M.A., Cotran, R.S., Leapman, S.B. and Folkman, J., 1974, Tumor
growth and neovascularization: An experimental model using the rabbit
cornea, J.N.C.I., 52:413.
Greenberg, H.S., Ensminger, W.D., Chandler, W.F., Layton, P.B., Junck, L.,
Knake, J. and Vine, A.D., 1984, Intra-arterial BCNU chemotherapy for
treatment of malignant gliomas of the central nervous system,
J.Neurosurg., 61:423.
Greig, N.H., 1987, Optimizing drug delivery to brain tumors, Cancer
Treat.Rev., 14:1.

Harbaugh, R.E., Saunders, R.L. and Reeder, R., 1988, Use of implantable pumps for central nervous system drug infusions to treat neurological disease, Neurosurgery, 23:693.

Hochberg, F.H. and Pruitt, A., 1980, Assumptions in the radiotherapy of glioblastoma, Neurology, 30:907.

Hanker, J.S. and Giammara, B.L., 1988, Biomaterials and biomedical devices, Science, 242:885.

Langer, R., 1983, Implantable controlled release systems, Pharmac.Ther., 21:35.

Langer, R., Brem, H. and Tapper, D., 1981, Biocompatibility of polymeric delivery systems for macromolecules, J.Biomed.Mat.Res., 15:267.

Langer, R. and Folkman, J., 1976, Polymers for the sustained release of proteins and other macromolecules, Nature, 263:797.

Langer, R. and Wise, D., eds., 1986, "Medical Applications of Controlled Release", Vols. I and II, CRC Press, Boca Raton, Florida.

Leong, K.W., D'Amore, P., Marletta, M. and Langer, R., 1986, Bioerodible polyanhydrides as drug-carrier matrices. II. Biocompatibility and chemical reactivity, J.Biomed.Mat.Res., 20:51.

Levin, V.A., Patlak, C.S. and Landahl, H.D., 1980, Heuristic modeling of drug delivery to malignant brain tumors. J.Pharmocokinetics and Biopharmaceutics, 8:257.

Loo, T.L. and Dion, R.L., 1965, Colorimetric method for the determination of 1,3-Bis(2-chloroethyl)-1-nitrosourea, J.Pharm.Sci., 54:809.

Neuwelt, E.A., Howieson, J., Frenkel, E.P., Specht, H.D., Weigel, R., Buchan, C.G. and Hill, S.A., 1986, Therapeutic efficacy of multiagent chemotherapy with drug delivery enhancement by blood-brain barrier modification in glioblastoma, Neurosurgery, 19:573.

Ringkjob, R., 1968, Treatment of intracranial gliomas and metastatic carcinomas by local application of cytostatic agents, Acta Neurol.Scand., 44:318.

Rosenblum, M.L., Bowie, D.L. and Walker, M.D., 1973, Diffusion in vitro and in vivo of 1-(2-chloroethyl)-3-(trans-4-methylcyclohexyl)-1-nitrosourea from silicone rubber capsules: A potentially new mode of chemotherapy administration, Cancer Res., 33:906.

Rubin, R.C., Ommaya, A.K., Henderson, E.S., Bering, E.A. and Rall, D.P., 1966, Cerebrospinal fluid perfusion for central nervous system neoplasms, Neurology, 16:680.

Tamargo, R.J., Epstein, J.I., Reinhard, C.S., Chasin, M. and Brem, H., 1989, Brain biocompatibility of a biodegradable controlled release polymer in rats, J.Biomed.Mat.Res., 23:253.

Tamargo, R.J., Epstein, J.I., Reinhard, C.S., Chasin, M. and Brem, H., 1988, Brain biocompatibility of a biodegradable polymer capable of sustained release of macromolecules, American Association of Neurological Surgeons, Toronto, p. 399.

Tamargo, R.J., Myseros, J.S. and Brem, H., 1988, Growth inhibition of the 9L gliosarcoma by the local sustained release of BCNU: a comparison of systemic versus regional chemotherapy, American Association of Neurological Surgeons, Toronto, p. 212.

Warnke, P.C., Blasberg, R.G. and Groothuis, D.R., 1987, The effect of hyperosmotic blood-brain barrier disruption on blood-to-tissue transport in ENU-induced gliomas, Ann.Neurol. 22:300.

Weiss, S.R. and Raskind, R., 1969, Treatment of malignant brain tumors by local methotrexate, International Surg., 51:149.

Yang, M.B., Tamargo, R.J. and Brem, H., 1989, Controlled delivery of 1,3-Bis(2-chloroethyl)-1-nitrosourea from ethylene-vinyl acetate copolymer, Cancer Res., 49:5103.

POLYMERIC DELIVERY SYSTEMS

Robert Langer

Massachusetts Institute of Technology, Department of Applied
Biological Sciences, Cambridge, MA 02139, USA

INTRODUCTION

Over the past decade there has been increasing attention devoted to the
development of controlled release systems for drugs, pesticides, nutrients,
agricultural products and fragrances. However, nearly all of the systems
that have been developed have not been capable of slowly releasing drugs of
large molecular weight (MW>600). In fact, up until 1976 it was a fairly
common conception in the field of controlled release that effective systems
could not be developed for macromolecules. However, after several years of
effort we discovered an approach that permitted the continuous release of
biologically active macromolecules as large as 2,000,000 daltons from
normally impermeable, yet biocompatible, polymers for over 100 days (Langer
and Folkman, 1976). In this paper we review two areas of our research
1) systems that release large molecules through porous polymer matrices and
2) modulated controlled release polymer systems.

POROUS DELIVERY SYSTEMS FOR THE RELEASE OF PROTEINS AND MACROMOLECULES

Our interest in creating controlled release systems for polypeptides
and other macromolecules began in 1974 and stemmed from studies on the
growth of solid tumors. Most solid tumors require ingrowth of new blood
vessels from the host for further tumor development and we were attempting
to isolate a drug that prevents the growth of new blood vessels. This
substance is derived from cartilage, a tissue that contains no blood
vessels. The bioassay used for this substance involved placing a tumor in
the cornea of a rabbit and monitoring the growth of new vessels toward the
tumor. We wanted to deliver the drug to the tumor to see if it decreased
the rate of blood vessel growth. The assay takes 30 days.

Purified fractions of the cartilage material were highly soluble, so
that they disappeared quickly after they were added. Therefore, we needed a
small sustained-release system to provide steady diffusion into the tumor.
Such a system had to be inert and noninflammatory. In early work (Gimbrone
et al, 1974), polyacrylamide pellets had been tried for this purpose. The
test protein was mixed with acrylamide before polymerization. After polymerization, however, the small pellets were often highly inflammatory. The
inflammation could be reduced by extensive washing, but it could never be

completely eliminated. Furthermore, washing leached out most of the test protein.

The only polymer systems reported for administering large molecules were those described by Davis (1974), namely polyacrylamide or polyvinyl-pyrrolidone. However, these systems damaged the cornea and permitted only brief periods of sustained release (Langer et al, 1981; Langer and Folkman, 1976). Therefore, we looked for other polymers and new ways of placing drugs in these polymers. However, one problem we found was that large molecules would only diffuse through highly porous and permeable membranes such as millipore filters. In these cases diffusion was too rapid to be of value. We thus developed a new approach that permitted sustained release of large molecules from biocompatible polymers (Langer and Folkman, 1976). We dissolved the polymer in an appropriate solvent and added the macromolecule in powder form. The resuting mixture can be cast in a mold and dried. When the pellets are placed in water, they release the molecules trapped within the polymer matrix.

We tested a number of polymer systems for tissue biocompatibility and release kinetics. Our best long-term release results were obtained with hydrophobic polymers. Examples included ethylene-vinyl acetate or poly-lactic acid. Certain hydrogels like polyhydroxyethylmethacrylate or poly-vinylacohol also worked effectively, but released proteins for shorter time periods. With the hydrophobic polymers, biologically active protein was released for over 100 days (Langer and Folkman, 1976). In other tests larger molecules (2 million mol wt) such as polysaccharides and polynucleo-tides, were also successfully released for long time periods (Langer and Folkman, 1976).

While these initial studies demonstrated the feasibility of releasing macromolecules from biocompatible polymers the kinetics were often not reproducible; we had not achieved controlled release. The irreproducibility resulted from drug settling and redistribution during casting and drying, caused by the insolubility of the incorporated macromolecule powder in the polymer solvent. At room temperature, the drug migrated vertically and visible lateral motion was caused by currents (possibly thermal) in the mixture. We thus developed a low-temperature casting and drying procedure to minimize this drug movement during matrix formation. By casting the dissolved polymer-solid drug powder mixture in a mold at −80°C, the entire matrix froze before any settling could occur. These matrices were then dried at −20°C for 2 days until almost all the solvent was gone. Final drying was conducted under vacuum at room temperature (Rhine et al, 1980).

Factors Affecting Release Kinetics

With this reproducible method, we could now accurately assess factors that regulated release kinetics. We found that drug powder particle size and drug loading (drug:polymer ratio) influence release time (Rhine et al, 1980). We coated drug-containing polymeric matrices by dropping each matrix into polymer solutions of differing concentrations. The coated matrices were dried and tested for release kinetics. We found that an increase in coating solution concentrations significantly decreases release kinetics. By combining these simple fabrication parameters (drug particle size, loading, and coating) release rates for any drug could be changed several thousandfold (Rhine et al, 1980).

To understand the release mechanism, cryomicrotomy was used to slice 10 μm-thick sections throughout the matrices. Viewed under an optical microscope, polymer films cast without proteins appeared as non-porous sheets. Matrices cast with proteins and sectioned prior to release dis-played areas of either polymer or protein. Matrices initially cast with proteins and released to exhaustion (5 months) appeared as porous films. Pores with diameters as large as 100 μm, the size of the protein particles, were observed. The structures visualized were also confirmed by Nomarski (differential interference contrast micrscopy). It appears that although pure polymer films were impermeable to macromolecules (Langer and Folkman, 1976), molecules incorporated in the matrix dissolve once water penetrates the matrix and are then able to diffuse to the surface through pores created as the particles of molecules dissolve. Scanning electron microscopy shows that the pores are interconnected.

Next investigated were changes in pore structure over time. Sections were prepared from matrices in the process of release. We observed that 1) the pore structure changes minimally as function of time, 2) after 16 or 40 hours there is no evidence of a receding interface between dissolved and dispersed drug, and 3) none of the drug remains undissolved at 40 h (30% release). Observations 2) and 3) differ from those reported for less soluble low molecular weight drugs such as certain steroids, and are probably due to the high solubilities of many proteins such as bovine serum albumin (BSA) (solubility >500 mg/ml). Thus, the widely used moving zone models developed by the Higuchis will not be applicable to the situation of macromolecules because of observations 2) and 3).

We made and then verified a number of assumptions to develop a model: 1) the rate limiting step for transport is drug diffusion through pores (other steps such as water penetration into the matrix and drug dissolution occur in less than 40 h); 2) The effect of concentration dependence on the drug diffusion coefficient is not significant. This was verified by an analysis of diffusion effects at the concentrations in the matrix; 3) No drug diffusion occurs through the polymer backbone (Langer and Folkman, 1976); 4) The pores are interconnected, the porosity is uniform, and pore size changes minimally with time; 5) The initial drug distribution is uniform. This was also verified by cryomicrotomy as discussed later; 6) No boundary layer effects exist. This was verified by stirring, which would have disrupted boundary layers had they been present. Release rates of slabs stirred in containers at 2000 rpm were identical to unstirred release rates, indicating the lack of boundary layer effects; 7) Infinite sink conditions exist. The volume of the release medium is approximately 100 times the volume of the polymer/protein matrix. Increasing the volume does not alter measured release kinetics; 8) Minimal effects exist due to osmosis or charge interaction of the drug with the polymer. Consonant with this assumption, we found no effect on release rate due to increasing the ionic strength of the medium from 0 to 1M NaCl.

For these assumptions, permitting release from only one side of the slab, the boundary conditions are those of zero flux at the coated edges, and C=0 at the releasing face.

If diffusion through pores occurs, we can solve the Fick diffusion equation

$$\delta C/\delta t = D_e \, \delta C/\delta x^2 \qquad (1)$$

The solution for this case is:

$$M_t/M_\infty = 1-(8/\pi^2) \sum^{\infty} \exp[-D_e(2n+1)^2\pi^2 t/4L^2]/(2n+1)^2, \qquad (2)$$
$$n = 1$$

where M_t is the cumulative drug mass released, M is the drug mass originally incorporated in the matrix, t is time (h) and L is the thickness of the slab (cm). In addition, we will set D_e, the effective diffusion coefficient

(cm^2/h) of the drug in the matrix, equal to $D_0 F$, where D_0 is the bulk diffusion coefficient of the same drug in the release media that has filled the pores and F is a factor accounting for the geometric effects of the pore structure of the matrix (i.e. tortuosity, dead-end pores, and constrictions between pores). F was determined via a regression analysis for several cases of BSA released from polymer slabs at several porosities. We found that a log-log plot of F versus porosity was well fit by the function

$$\log_{10} F = 0.463 + 5.64 \log_{10} e \tag{3}$$

where e is the porosity. Knowing this equation for F, we can then write

$$D_e = D_0 (2.904 \, e^{5.64}) \tag{4}$$

and this value of D_e can be substituted into equation 2.

Both the slab thickness L and the porosity e were measured. For a given macromolecule, the bulk diffusivity, D_0, is measurable or obtainable from the literature. Thus, a test of the model is to cast slabs using other proteins, measure the parameters L, e, and D_0, and see whether the release kinetics follow equation 2. This has been done for β-lactoglobulin and lysozyme and predictions based on equation 2 show general agreement with the data obtained (Balazs et al, 1985).

An additional check of the model is to determine if it can predict not only the time-dependent release of the drug, but the time-dependent position of the drug within the matrix. If equation 2 is valid, then the drug distribution within the matrix can be described by:

$$C(x,t) = (4C_0/\pi) \sum_{n=1}^{\infty} [(-1)^n/(2n+1)] \exp[-D_e(2n+1)^2 \pi^2 t/4L^2] \cos[(2n+1)\pi x/2L].$$

$$n = 1 \tag{5}$$

where C_0 is the initial concentration of drug in the matrix [mg/(cm^3 matrix)], and $C(x,t)$ is the concentration (the concentration C and C_0 are expressed in terms of the volume of the whole matrix, including both pore and polymer volumes) at time t and distance x (cm) into the matrix from the exposed face. To test eq. 5, we used cryomicrotomy to determine the drug concentration profiles within the polymer matrix for several cases of loading and release time. Partially released matrices were sectioned at 10 μm intervals and the remaining protein in each section was assayed and plotted against its normalized (x/L) position within the matrix to yield internal concentration profiles. The data and the predictions from equation 5 are within experimental error (Balazs et al, 1985).

The diffusion equations used above are simplifications of more complex processes. The F factor was empirically derived and must take into account those matrix pore geometric factors contributing to decreases in diffusion rates. Such factors may include pore "tortuosity", dead-end pores, and pore constrictions. Initial modeling studies suggest that constrictions, in particular, have large effects in retarding release (Balazs et al, 1985; Siegel and Langer, 1986).

Comparison of In Vivo and In Vitro Release Kinetics and Biocompatibility

In vitro and in vivo release kinetics were compared using two different approcahes. In the first approach (the recovery approach) polymer implants containing a radioactively labelled substrate (^{14}C-BSA, ^{14}C-β-lactoglobulin, or ^{3}H-inulin) were implanted subcutaneously into rats (in vivo) or released in phosphate buffered saline, pH 7.4, at 37°C (in vitro). At various time points, the polymer implants were removed from the rats or the saline. They were then lyophilized to remove residual water and dissolved in xylene (xylene was chosen because it is miscible with the scintillation fluid). When the polymer dissolved, the unreleased macromolecules precipitated to the bottom of the vial. Water was then added to dissolve the macromolecules; the scintillation fluid was next added, resulting in a homogeneous translucent emulsion which was counted by liquid scintillation counting.

Release rates determined in this manner were essentially identical in the in vivo and in vitro implants. In addition, for the in vitro experiments, release was also measured directly by analyzing the radioactivity in the release media. The release rates determined in this way correlated precisely with the in vitro and in vivo release rates determined by the recovery experiments (see above). Furthermore, they demonstrated that the material balance was completed showing that no material was lost in the experiment (Brown et al, 1983).

One limitation of the above approach, however, was that we could not assay the amount of macromolecules directly released in vivo. This is because macromolecules such as proteins are metabolized, making direct in vivo release measurements difficult. To solve this problem we used ^{3}H-inulin as a model.

Inulin is a polysaccharide of molecular weight 5200. It is one of the very few molecules that is not metabolized in vivo nor reabsorbed or secreted by kidney tubules. Thus, all inulin released from the polymer should be recovered in the urine. An in vivo-in vitro comparison was made by making nine identical inulin-polymer pellets. Five pellets were implanted in rats housed in metabolic cages. Four pellets were released into a physiological solution of phosphate-buffered saline (PBS) at pH 7.4 at 37°C. Both urine and PBS were collected daily. The ^{3}H-inulin was measured by scintillation counting. The experiment was carried out for 500 h. (Additional experiments have been carried out for 1500 h with similar results.) Over this period, in vivo and in vitro release rates agreed to within 1% (Brown et al, 1983).

Furthermore, as an internal control, inulin pellets were removed from several animals at 450 h and the urine analyzed 4.5 h later. Within that time the inulin recovery rate had dropped by a factor of over 50.

The polymer slabs were examined histologically in two different in vivo sites at times as long as seven months after implantation. Nearly no inflammation or fibrous encapsulation was observed (Brown et al, 1983).

Thus, these experiments show that in vitro and in vivo release kinetics of macromolecules from ethylene-vinyl acetate copolymer slabs are essentially identical and they establish a methodology which can be easily applied for future in vitro-in vivo release comparisons.

Approaches for Achieving Zero-Order Release

One of our goals was the development of a zero-order release device for macromolecules. The difficulty is that our systems contain drug evenly distributed through polymer slabs and, thus, release rates will decrease with time because the drug diffusion distance from the matrix surface increases with time. In order to obtain zero-order release, we felt it would be necessary to either compensate for the distance dependent diffusion in a matrix device or to employ a different kind of release system such as a reservoir device (a system in which all drug is centered inside a membrane). Variations of the latter approach have been reported for low molecular weight drugs in which matrices have been laminated with rate controlled outer barriers. However, we felt that such an approach would prove difficult for macromolecules because they would decrease, if not eliminate, the permeability of these large molecular weight drugs. We therefore focused on ways of compensating for the distance dependent diffusion. The approaches we considered were altering the drug distribution pattern or matrix geometry. We felt the idea of altering matrix geometry would be experimentally more attractive, particularly if we would be able to design a shape in which an increased area of drug would be available as the diffusion distance increased.

We therefore developed models for predicting release from different shaped systems and found that a special hemisphere system (coated everywhere with an impermeable coating except for an aperture in the center face) worked best. We then developed an experimental method for making the special hemisphere device to test our predictions. In this procedure we cut off the bottoms of test tubes and used them as molds. A low temperature casting and drying procedure was used. We were able to use this approach to obtain a constant release rate of 0.5 mg/day for two months (the duration of the experiment) for test macromolecules (Hsieh et al, 1983).

New Formulation Procedures

A number of new formulation procedures have been developed. The first of these does not require solvents. The method consists of mixing drug and polymer powders below the glass transition temperature of the polymer and compressing the mixture at a temperature above the glass transition point. The macromolecule is not exposed to any organic solvent during the fabrication process.

The advantages of this sintering method, when compared to solvent casting methods used in previous studies, include: 1) elimination of shrinkage, 2) elimination of the need for potentially expensive scale-up steps such as vacuum drying, 3) reduction of processing time (slabs have been produced in two h, compared to four days required for solvent casting), and 4) lack of the necessity to expose the macromolecule to a solvent. The kinetics, microstructure, bioactivity and effects of polymer and drug particle size have all been studied (Cohen et al, 1984).

A simple technique has also been devised for preparing microspheres containing macromolecular drugs. Beads with good sphericity were formed by a simple extrusion process that can easily be repeated by any laboratory without special equipment. Drug release kinetics and polymer microstructure were also investigated with this technique. The kinetic trends were as would be predicted from our earlier studies (Sefton et al, 1984).

A technique for insuring the controlled release of small amounts of macromolecules such as polypeptides from polymeric delivery systems was also developed. We found that incorporation of albumin in milligram quantities into these delivery systems can facilitate the sustained release of nanogram or microgram quantities of model macromolecules such as epidermal growth factor (EGF) (Murray et al, 1983).

Applications

Studies have also been conducted to explore numerous applications of these systems. These include release of insulin (Brown et al, 1986), anticalcification agents (Levy et al, 1985), interferons (Langer et al, 1981, growth factors (Langer and Folkman, 1978) and inhibitors (Lee and Langer, 1983), and neurologically active agents (During et al, 1990; Mayberg et al, 1981).

MODULATED RELEASE SYSTEMS: MAGNETIC SYSTEMS

Several polymeric systems capable of delivering drugs at increased rates on demand were studied and developed. The first system consists of drug powder dispersed within a polymeric matrix (generally ethylene vinyl acetate copolymer, EVAc), together with magnetic beads.

Release rates were controlled by an oscillating external magnetic field, which is generated by a device that rotates permanent magnets beneath the vials. By placing small plastic cages containing animals on the top disc, it can also be used for in vivo studies. Polymer matrices containing drug and magnets can release up to 30 times more drug when exposed to the magnetic field, and release rates return to normal when the magnetic field is discontinued. The magnetically controlled implant does not cause inflammation in vivo. This was confirmed by the lack of edema, cellular infiltrate or neovascularization as judged by gross and histologic examination in animals (Hsieh et al, 1981).

Magnetic Field Characteristics

The amplitude of the magnetic field was varied by increasing the distance between the external and embedded magnets or by changing the embedded magnet's strength. The extent of release enhancement increases as the field amplitude rises. For example, at a field frequency of 11 Hz, a mean release rate enhancement of 12.4 times was obtained for matrices containing 1100 G magnets, as opposed to 1.5 times for matrices containing less than 100 G magnets (Edelman et al, 1985).

When the frequency of the applied field was increased from 5.0 Hz to 11.0 Hz the rate of release rose linearly (Edelman et al, 1985).

To study magnet orientation, samples containing a single 1100 G magnet were used. In half the cases, the magnet was placed perpendicular to the applied field and in the other half it was oriented parallel to the field. The mean release rate enhancement was 2.1 times in the parallel cases, and 12.4 times in the perpendicular cases. The difference between the two is presumably due to rotational torque. When placed in parallel the magnet rotates in an attempt to align its pole vector with the field; therefore the displacement is smaller (Edelman et al, 1985).

Polymer Properties

The mechanical properties of the polymeric matrix also affect the extent of magnetic enhancement. For example, the modulus of elasticity of the polymeric matrix can be altered by changing the vinyl acetate content of the polymer. The release rate enhancement induced by the magnetic field increases as the modulus of elasticity of the EVAc decreases (Kost et al, 1985).

Mechanism

The release of macromolecules from EVAc systems not containing magnetic beads has been studied extensively. It was found that although molecules with molecular weights greater than 300 cannot permeate through the polymer, the direct incorporation of macromolecules in the polymer-macromolecule casting procedure caused a tortuous and complex series of pores to form within the matrix. Factors affecting permeation of water into the polymer and drug out of these pores determine the release rates.

Video recordings of the polymer matrix surface show that the beads actually move within the matrix in response to the external magnetic field and move adjacent material containing polymer and drug with it, "squeezing" out the dissolved drug through the pores (Edelman et al, 1985).

McCarty et al (1984) proposed a model for the enhanced release and suggested that the major effect stems from the alternate compression and expansion deformation of the pores, causing the fluid within to undergo a pulsatile flow which alone (no net convection) is able to greatly improve diffusive mass transfer. A mathematical model was developed based on ideal axial diffusion in a cylinder with pulsed flow, which describes the observed trend of increasing release rate with increasing frequency. This provides a preliminary estimate of drug release rate enhancement as a function of frequency of the oscillating magnetic field (McCarty et al, 1984).

Release Dynamics

A flow-through spectrophotometer provided a continuous recording reflecting BSA release. Release rates were very stable at baseline. When a matrix was exposed to a 390-G magnetic field for 20 min the absorbance of BSA rose after the magnetic field was applied, plateaued at this elevated level, and then returned to baseline after the field was withdrawn. Within the time constants of the system the rise and fall were instantaneous, and the increased release remained at a constant elevated level for the duration of field exposure (Edelman et al, 1987).

Refractory Time

The modulation (release rate during magnetic exposure compared to release rate without magnet exposure) was independent of the duration of the interval between repeated pulses. The same modulation was elicited for the same strength magnetic fields regardless of the time between application of the pulses (Edelman et al, 1987).

Field Amplitude

When the strength of the magnetic field was altered by adjusting the input voltage to an electromagnet, modulation changed accordingly. The relationship between modulation and the strength of the applied magnetic fields ranging from 200 to 700 G is linear (Edelman et al, 1987).

Controls

Heat had little effect on the rate of BSA release. The rate of BSA release was nearly constant from polymer matrices transferred every two h from 25 to 37°C. The heat generated by the electromagnet did not alter release rates from polymer matrices either. The temperature rose at the electromagnet surface at a rate of 8°C/h for the first 30 min after the magnet was turned on. The rate of BSA release from a polymer matrix without an embedded magnet adjacent to the electromagnet face did not change over this time period and for seven h afterward (Edelman et al, 1987).

In Vivo Studies

Using insulin as a marker the above effects observed in vitro (response time, response duration, amplitude) were also observed in vivo (Edelman et al, 1987).

Furthermore, implants composed of EVAc-embedded magnets and bovine zinc insulin were placed s.c. in diabetic rats for two months. After implantation the blood glucose level decreased due to diffusion of insulin from the polymer. When the diabetic rats were exposed to an oscillating magnetic field, the blood glucose levels were further lowered from 50 to 200 mg dl^{-1}

below this basal level depending on the magnetic field conditions. These
results were confirmed by radioimmunoassays. This phenomenon was not observed in (i) diabetic rats receiving EVAc implants with insulin without a
magnet, (ii) control rats receiving implants with a magnet but without
insulin, (iii) diabetic rats not containing any implant. All these animals
were exposed to the same manipulations as the experimental groups of animals
(Edelman et al, 1987).

In addition to these approaches we have also found that ultrasound
(Kost et al, 1986) can enhance transport rates through both degradable and
non-degradable polymers, and that enzymes (Kost et al, 1988) can be incorporated into polymers to provide direct feedback control.

REFERENCES

Balazs, A.C., Calef, D.F., Deutch, J.M., Siegel, R.A. and Langer, R., 1985,
    The role of polymer matrix structure and interparticle interactions
    in diffusion limited drug release, Biophys.J., 47:97.
Brown, L., Siemer, L., Munoz, C., Edelman, E. and Langer, R., 1986,
    Controlled release of insulin from polymer matrices: control of
    diabetes in rats, Diabetes, 35:692.
Brown, L., Wei, C. and Langer, R., 1983, In vitro and in vivo release of
    macromolecules from polymeric drug delivery systems, J.Pharm.Sci.,
    72:1181.
Cohen, J., Siegel, R. and Langer, R., 1984: Sintering technique for prepar-
    ation of polymer matrices for the sustained release of macromole-
    cules, J.Pharm.Sci., 73:1034.
Davis, B.K., 1974, Diffusion in polymer gel implants, Proc.Natl.Acad.Sci.,
    USA, 71:3120.
During, M.J., Freese, A., Sabel, B.A., Saltzman, W.M., Deutch, A., Roth,
    R.H. and Langer, R., Controlled release of dopamine from a polymeric
    brain implant: in vivo characterization, Ann.Neur., in press.
Edelman, E., Brown, L. and Langer, R., 1987, Magnetic controlled release
    system in vitro and in vivo, J.Biomed.Mat.Sci., 21:339.
Edelman, E., Kost, J., Bobeck, H. and Langer, R., 1985, Regulation of drug
    release from porous polymer matrices by oscillating magnetic fields,
    J.Biomed.Mat.Res., 19:67.
Gimbrone, M.A., Jr., Contran, R.S., Leapman, S.B. and Folkman, J., 1974,
    J.Natl.Cancer Inst., 52:413.
Hsieh, D.S.T., Langer, R. and Folkman, J., 1981, Magnetic modulation of
    release of macromolecules from polymers, Proc.Natl.Acad.Sci.USA,
    78:1863.
Hsieh, D., Rhine, W. and Langer, R., 1983, Zero-order controlled release
    polymer matrices for micromolecules and macromolecules, J.Pharm.
    Sci., 72:17.
Kost, J., Leong, K. and Langer, R., 1986, Ultrasonic modulated drug delivery
    systems in polymers, in: "Medicine, Biomedical and Pharmaceutical
    Applications" II, E. Chiellini, ed., Plenum Press.
Kost, J., Leong, K. and Langer, R., 1988, Ultrasonically controlled poly-
    meric drug delivery, Makromol.Chem.Macromol.Symp., 19:275.
Kost, J., Noekker, R., Kunica, E. and Langer, R., 1985, Magnetically con-
    trolled release systems: effect of polymer compositon, J.Biomed.Mat.
    Res., 19:935.
Langer, R., Brem, H. and Tapper, D., 1981, Biocompatibility of polymeric
    delivery systems for macromolecules, J.Biomed.Mat.Res., 15:267.
Langer, R. and Folkman, J., 1976, Polymers for the sustained release of pro-
    teins and other macromolecules, Nature, 263:797.
Langer, R.S. and Folkman, J., 1978, Sustained release of macromolecules from
    polymers, in: "Polymeric Delivery Systems", Midland Macromolecular
    Monograph, 5:175, R.J. Kostelnik, ed., Gordon and Breach, New York.

Langer, R., Hsieh, D.S.T., Brown, L. and Rhine, W., 1981, Polymers for the sustained release of macromolecules: controlled and magnetically modulated systems, in: "Better Therapy with Existing Drugs: New Uses and Delivery Systems", A. Bearn, ed., Merck & Company, Biomedical Information Corporation, New York.

Lee, A. and Langer, R., 1983, Shark cartilage contains inhibitors of tumor angiogenesis, Science, 221:1185.

Mayberg, M., Langer, R., Zervas, N. and Moskowitz, M., 1981, Perivascular meningeal projections from cat trigeminal ganglia: possible pathway for vascular headaches in man, Science, 213:228.

McCarty, M., Soong, D. and Edelman, E., 1984, Control of drug release from polymer matrices impregnated with magnetic beads – a proposed mechanism and model for enhanced release, J. of Controlled Release, 1:143.

Murray, J., Brown, L., Klagsbrun, M. and Langer, R., 1983, A micro sustained release system for epidermal growth factor, In Vitro, 19:743.

Rhine, W., Hsieh, D.S.T. and Langer, R., 1980, Polymers for sustained macromolecule release: procedures to fabricate reproducible delivery systems and control release kinetics, J.Pharm.Sci., 69:265.

Sefton, M.V., Brown, L.R. and Langer, R., 1984, Ethylene-vinyl acetate microspheres for controlled release of macromolecules, J.Pharm.Sci., 73:1859.

Siegel, R.A. and Langer, R., 1986, A new Monte Carlo approach to diffusion in constricted porous geometries, J.Colloid Interfacial Science, 109:426.

Participants of the NATO Advanced Studies Institute "Vaccines: Recent Trends and Progress" held at Cape Sounion Beach, Greece during 24 June-5 July 1990. The organizing committee included A.C. Allison (ASI Co-Director), K. Dalsgaard, G. Gregoriadis (ASI Director and Chairman), G. Poste and H. Snippe.

CONTRIBUTORS

Ahlers, M., Institut für Organische Chemie, Universität Mainz, D-6500 Mainz West Germany

Aston, R., Peptide Technology Ltd., P.O. Box 444, Dee Why, NSW 2099, Australia

Blankenburg, R., Institut für Organische Chemie, Universität Mainz, D-6500 Mainz, West Germany

Bomford, R., Wellcome Biotechnology, Langley Court, Beckenham, Kent BR3 3BS UK

Brem, H., Departments of Neurological Surgery, Ophthalmology and Oncology, Johns Hopkins University School of Medicine, Baltimore, Maryland, USA

Cable, C., The Department of Pharmacy, University of Strathclyde, Glasgow UK

Cassidy, J., Department of Medical Oncology, University of Glasgow, Glasgow UK

Cobb, L., Division of Experimental Pathology and Therapeutics, Medical Research Council Radiobiology Unit, Harwell, Didcot, Oxon, OX11 0RD, UK

Engert, A., Drug Targeting Laboratory, Imperial Cancer Research Fund, Lincoln's Inn Fields, London WC2A 3PX, UK

Florence, A.T., The Centre for Drug Delivery Research, The School of Pharmacy, University of London, 29-39 Brunswick Square, Holborn, London, WC1N 1AX, UK

Gaber, B.P., Center for Bio/Molecular Science and Engineering, Code 6090, Naval Research Laboratory, Washington, DC 20375-5000, USA

Gabizon, A., Hadassah Medical Center, Jerusalem 91120, Israel

Goins, B., Center for Bio/Molecular Science and Engineering, Code 6090, Naval Research Laboratory, Washington, DC 20375-5000, USA

Grainger, D.W., Department of Chemical and Biological Sciences, Oregon Graduate Institute of Science and Technology, Beaverton, OR 97006-1999, USA

Kaye, S.B., Department of Medical Oncology, University of Glasgow, Glasgow UK

Kikukawa, A., Research Division, Green Cross Corporation, Shodai-ohtani 2-1180-1, Hirakata, Osaka 573, Japan

Langer, R., Massachusetts Institute of Technology, Department of Applied Biological Sciences, Cambridge, MA 02139, USA

Leserman, L., Centre d'Immunologie INSERM-CNRS de Marseille-Luminy, Case 906, 13288 Marseille Cedex 9, France

Machy, P., Centre d'Immunologie INSERM-CNRS de Marseille-Luminy, Case 906, 13288 Marseille Cedex 9, France

Markham, N.I., Academic Department of Surgery, Royal Free Hospital, Pond Street, Hampstead, London UK

Meller, P., Institut für Organische Chemie, Universität Mainz, D-6500 Mainz, West Germany

Moghimi, S.M., Department of Biochemistry, Charing Cross and Westminster Medical School, Fulham Palace Road, London W6 8RF, UK

Novell, J.R., Academic Department of Surgery, Royal Free Hospital, Pond Street, Hampstead, London, UK

Papahadjopoulos, D., Cancer Research Institute and Department of Pharmacology, University of California, San Francisco, California 94143

Patel, H.M., Department of Biochemistry, Charing Cross and Westminster Medical School, Fulham Palace Road, London W6 8RF, UK

Price, R.R., Center for Bio/Molecular Science and Engineering, Code 6090, Naval Research Laboratory, Washington, DC 20375-5000, USA

Reichert, A., Institut für Organische Chemie, Universität Mainz, D-6500 Mainz, West Germany

Ringsdorf, H., Institut für Organische Chemie, Universität Mainz, D-6500 Mainz, West Germany

Rudolph, A.S., Center for Bio/Molecular Science and Engineering, Code 6090, Naval Research Laboratory, Washington, DC 20375-5000, USA

Salesse, C., Centre de Recherche en Photobiophysique, Université Quebec, Trois Rivieres, Quebec, Canada

Singh, A., Center for Bio/Molecular Science and Engineering, Code 6090, Naval Research Laboratory, Washington, DC 20375-5000, USA

Soria, M., Biotechnological Research Farmitalia Carlo Erba, 24 Viale E. Bezzi, Milano and Institute of Pharmacological Sciences, University of Milano, Italy

Thorpe, P., Drug Targeting Laboratory, Imperial Cancer Research Fund, Lincoln's Inn Fields, London WC2A 3PX, UK

Tomlinson, E., Advanced Drug Delivery Research, Ciba-Geigy Pharmaceuticals, Horsham, West Sussex, EH12 4AB, UK

Ueda, Y., Research Division, Green Cross Corporation, Shodai-ohtani 2-1180-1, Hirakata, Osaka 573, Japan

Yamanouchi, K. Research Division, Green Cross Corporation, Shodai-ohtani 2-1180-1, Hirakata, Osaka 573, Japan

Yokoyama, K., Research Division, Green Cross Corporation, Shodai-ohtani
2-1180-1, Hirakata, Osaka 573, Japan

INDEX

Polyethylene glycol
  protein conjugates, 8

Polymerized liposomes
  morphology, 107
  solvents, effect of, 107
  temperature, effect of, 107

Polymers
  biocompatible, 166
  in drug delivery, 156

Polyols
  liposomal stability, effect on,
    104–106

Polypeptides, 2
  half-life, 2

Prostaglandin E$_1$
  formulations, 137
  microspheres, 141
  pharmacological studies, 141

Proteins, 2
  biological disposition, 15
  chimaeric, 24
  in drug delivery, 165
  endocrine, 2
  extravasation of, 2
  half-life, 2
  hybrid, 13
  PEG-modified, 9
  therapeutic, 9, 12

Recombinant conjugates, 21–25

Reticuloendothelial system
  blockade of, 87
  liposomes, uptake of, 87

Sendai virus
  fusion with, 24

Serum opsonins
  freezing, effect of, 90
  heat, effect of, 90
  thawing, effect of, 90

Site-directed mutagenesis
  plasminogen activators, 10

Site-specific delivery, 3
  constraints, 3
  opportunities, 3

Site-specific systems, 3
  extravasation, 3

Streptavidin
  fluorescent, 59
  in monolayers, 59
  structure of, 67

Targeting
  by domain engineering, 23
  to solid tumours, 7
  strategies, 57, 58

Thyroid stimulating hormone
  activity, enhancement of, 54

Tissue-type plasminogen activator,
    10, 11
  deletion mutants of, 10

Tubules
  for anti-fouling agents, 111
  for controlled release, 111
  copper coated, 111
  diacetylene, 110
  morphology, 111
  tetracyclin-loaded, 112

Tumour-associated antigens, 29, 127

Tumours
  antigen distribution, 34, 35
  extravascular, 7
  drug delivery to, 7
  histology, 29, 31
  intratumour pressure, 34
  liposomes, uptake of, 98
  vascular architecture, 29

Vascular architecture
  carcinomas, 31–32
  normal tissue, 30
  sarcoma, 32, 33